MW00998567

The ASME® Code Simplified

Other Books and Handbooks of Interest from McGraw-Hill

The ASME® Code Simplified

Power Boilers

Dyer E. Carroll

Dyer E. Carroll, Jr.

McGraw-Hill

New York San Francisco Washington, D.C. Auckland Bogotá
Caracas Lisbon London Madrid Mexico City Milan
Montreal New Delhi San Juan Singapore
Sydney Tokyo Toronto

Library of Congress Cataloging-in-Publication Data

Carroll, Dyer E., date.
The ASME code simplified : power boilers / Dyer E. Carroll, Dyer E. Carroll, Jr.
p. cm.
Includes index.
ISBN 0-07-011636-9 (acid-free paper)
1. Steam-boilers—Standards—United States. I. Carroll, Dyer E., date. II. Title.
TJ288.C34 1997
621.1′83′021873—dc20 96-2856
CIP

McGraw-Hill

A Division of The ***McGraw-Hill*** *Companies*

1 2 3 4 5 6 7 8 9 0 DOC/DOC 9 0 1 0 9 8 7 6

ISBN 0-07-011636-9

The sponsoring editor for this book was Robert Esposito and the production supervisor was Pamela A. Pelton. It was set in Century Schoolbook by North Market Street Graphics.

Printed and bound by R. R. Donnelley & Sons Company.

This book is printed on recycled, acid-free paper containing a minimum of 50% recycled de-inked fiber.

This book is dedicated to
Betty W. Carroll
by her husband and son

Contents

Preface

This book is intended for manufacturers, designers, repairers, and inspectors of boilers designed in accordance with the requirements of Section I of the American Society of Mechanical Engineers Boiler & Pressure Vessel Code. Such boilers include steam boilers over 15 psig, high temperature hot water boilers (over 160 psig and/or over 250°F), miniature boilers, electric boilers, and organic fluid vaporizers. Such boilers must, as a matter of law, be designed, constructed, and inspected in accordance with Section I if they are to be installed in most of the 50 states of the United States or any of the 12 provinces of Canada. The states of Mexico are in the process of becoming code states or "jurisdictions" as well. Additionally, ASME code construction is widely used as a contractual requirement elsewhere in the world.

This book will help to clarify Section I of the Code, but it cannot substitute for the actual Code itself. Furthermore, boiler work requires that you have for reference current editions of the following:

Section I	Power Boilers
Section II, Parts A, D	Material Properties
Section V	Nondestructive Testing
Section IX	Welding and Brazing
ANSI/ASME B31.1	Power Piping
ANSI/ASME	National Board Inspection Code (for repair work)

The Boiler Code was first issued in 1914. Since 1953, the Code has been issued in a new edition every three years, with interim changes being published as addenda. The Code is now published in loose-leaf format and dated July 1 of the year of issue (e.g., July 1, 1995 for the '95 edition of the Code). Addenda are published as colored pages to be substituted for the original white pages. The addenda are published annually on December 31 of the year of issue. The addenda become mandatory six months after the date of issue except for boilers or pressure vessels contracted for before the end of the six-month period.

The Code books mentioned previously may be obtained from:

The American Society of Mechanical Engineers
United Engineering Center
345 East 47th Street
New York, New York 10017

Acknowledgments

The authors wish to thank the people who helped with the preparation of this book in various ways. Thanks to John Lynch of Hodge Boiler Works, Jackie Bunting of Clark-Reliance Corporation, Vin Helfrich of Helfrich Brothers Boiler Works, Phil Stillitano of Babcock & Wilcox, Richard Cyranowski of ABB-CE Services, Dick Helfrich of H & H Engineering, Lori Alten of Diamond Power Specialty Company, Chuck Quirk of Copes-Vulcan, Inc., David Lohr of USS Tubular Products, Raymond Swanson of the Uniform Boiler & Pressure Vessel Laws Society, Ivette Nieves of ASME, Elaine O'Neal of John Wiley & Sons, Inc., Sandra Sabino, Doug Hague of LR Insurance, Lee Ehrenzeller of the Frank I. Rounds Company, David Smith of Corenco, Marilyn Bourque of Crosby Valve and Gage Company, Edward Valves, Inc., Consolidated Industrial Valves, Cambridge Pipe, and Russ Huebner and Sharon Keefe of Carroll Engineers, Inc.

Introduction

Organization of the Code

At the present time the ASME Boiler & Pressure Vessel Code consists of 11 sections, but there are also subsections, divisions, and parts. The complete Boiler & Pressure Vessel Code appears in 25 volumes and costs over $5750.00, with the loose-leaf binders alone costing $500.00. Section I is the power boiler section, and henceforth any reference to "the Code" should be understood to mean Section I.

Section I consists of the Foreword, two statements of policy, the Preamble, nine parts, a mandatory Appendix I, and the Appendix, which contains explanations of the Code and is not mandatory unless specifically mentioned in one of the parts. It should be noted that there is a good deal of mandatory information in the Appendix. The Appendix to the Code also contains sample data report forms and guides.

At the end of Section I is an index. The various committee, subcommittee, subgroup, and working group titles and personnel are listed. However, technical inquiries must be addressed to:

Secretary
ASME Boiler and Pressure Vessel Committee
345 East 47th Street
New York, NY 10017.

Such inquiries must be prepared and submitted in accordance with the procedures described in the Code's Appendix I.

The Foreword is three pages in length. It discusses the Boiler and Pressure Vessel Committee and its function and objective. The manner of changing the Code is explained. The relationships between the Code committees and the various jurisdictions that enforce the Code as a matter of law and the composition of the National Board of Boiler and Pressure Vessel Inspectors, as well as the relationship of the National Board to the Code, are explained. The interchangeability of ASTM-specified and ASME-specified materials is discussed. The terms "Manufacturer" and "Authorized Inspector" are introduced.

The statements of policy discuss the use of Code symbols in advertising and the use of ASME markings to identify manufactured items. Briefly, both practices are encouraged, provided that there is no misrepresentation.

The Preamble defines the terms "power boiler," "electric boiler," "miniature boiler," and "high-temperature water boiler." The terms "boiler proper" and "boiler external piping" are defined, and the use of B31.1 for the materials, design, fabrication, installation, and testing of the boiler external piping is mandated. The terms "fired" and "unfired steam boiler" are defined. The use of the rules of Section VIII, Division 1, as an alternative to Section I rules for unfired steam boilers and for the expansion tanks required in connection with high-temperature water boilers is expressly permitted. The requirement that organic fluid vaporizers be designed in accordance with Section I is also given in the Preamble.

Part PG is about 73 pages in length and covers general requirements for all methods of construction. The service limitations (PG-2) are stated: Section I rules apply to boilers in which steam is generated at a pressure of more than 15 psig and to high-temperature water boilers intended for operation at pressures exceeding 160 psig and/or temperatures exceeding 250°F. Boilers for use below these service limits may be constructed in accordance with Section I rules but would normally be expected to be constructed in accordance with Section IV (i.e., be H-stamped).

Materials requirements are stated in PG-5 through PG-13. All pressure-retaining parts must be made of materials listed in Section II and meet one of the specifications listed in PG-6 (plate), PG-7 (forgings), PG-8 (castings), PG-9 (pipes, tubes, and pressure-containing parts), or PG-13 (stays). It is emphasized that only Section I materials can be used for boiler service. It should be noted that austenitic stainless steels are not permitted for wetted service generally (although there are a few specific exceptions to the prohibition); they may be used in superheaters.

The balance of Part PG (PG-16 to PG-113) contains rules for design, safety valves, fabrication, inspection, and stamping and data reports. These paragraphs will be the subjects of later chapters of this book.

Part PW is about 30 pages in length and covers requirements for boilers constructed by welding. This includes design requirements, nondestructive testing requirements, test plate testing, and postweld heat treatment requirements. The design requirements include weld size and strength specifications for reinforced openings. This part also will be the subject of a later discussion.

Part PR is five lines long and covers requirements for riveted boilers. The Code states that riveted boilers shall be constructed to the 1971 edition of Section I. We will follow the example of the Code and leave riveted construction in the past.

Part PWT is three pages long and covers requirements for watertube boilers. There is a table (Table PWT-10) that gives tube thicknesses required for different pressures for the most common boiler tube specifications and for metal temperatures not exceeding 700°F.

Part PFT is 20 pages long and covers firetube boiler requirements. This part contains rules for designing cylindrical parts exposed to external pressure and calculating stays and stayed surfaces, corrugated furnaces, and safety valve openings in shells of firetube boilers. Rules from this part are used in many types of problems described hereafter.

Part PFH is one page covering optional requirements (i.e., conditions under which Section VIII, Division 1 rules may be used) for the feed water heater.

Part PMB is three pages in length and covers requirements for miniature boilers. Miniature boilers are defined as those that do not exceed:

1. 16-inch inside shell diameter.
2. 20 square feet of heating surface.
3. 5 cubic feet gross volume.
4. 100 psig MAWP (Maximum Allowable Working Pressure—PMB-2).

These boilers may be of welded construction, but unlike other boilers, miniature boilers do not require radiography or postweld heat treatment of the welded joints (PMB-9). However, the miniature boilers must be hydrostatically tested at three times the MAWP (PMB-21), while other boilers are tested at 1.5 times the MAWP (PG-99).

Part PEB is four pages in length and covers requirements for electric boilers. Electric boilers may be built and stamped S or M, but there is also provision for use of an E stamp by a Manufacturer that does not have the S or M stamps. E stampholders are limited to construction methods not including welding or brazing (PEB-2.2).

The Preamble states: "Unfired steam boilers shall be constructed under the provisions of Section I or Section VIII"; PEB-3 states: "The boiler pressure vessel may be constructed in compliance with the ASME Pressure Vessel Code Section VIII, Division 1, rules for unfired steam boilers (UW-2(c)) subject to the following conditions." It goes on to say that the *boiler pressure vessel* may be built and stamped by a U stampholder. It should be noted and emphasized, however, that the completed electric boiler must be stamped with an S, M, or E stamp, as appropriate, even if the boiler pressure vessel is U-stamped. The U stampholder cannot build completed electric boilers. There must be an S, M, or E stampholder. (This is clear both from PEB-3.1 and from PEB-19.2.)

Part PVG is three pages long and covers requirements for organic fluid vaporizers. The major thrust of this part is spelled out in PVG-12, which covers special safety valve requirements and the use of rupture disks for protecting organic fluid vaporizers from overpressure while preventing or controlling discharge of vapor.

Appendix I is one page and describes the mandatory method and format for submitting technical inquiries to the Code committee. Failure to follow Appendix I will result in the inquiry being returned unanswered.

The Appendix is about 88 pages in length. It includes design rules, quality control system requirements, sample calculations of various types, and forms and guides to filling out the forms. In the chapters that follow, there will be numerous references to the Appendix.

The ASME® Code Simplified

Chapter

1

Development of Boilers, Materials, and Codes

The boiler and pressure vessel engineers active today have advantages not available to the pioneers in the development of boilers and pressure vessels. They have the advantages of being able to refer to established codes for design and construction of boilers. They have the advantage of picking up Section II of the ASME Code and selecting materials that are suitable for construction of boilers. They may use Section I to design and calculate the maximum allowable working pressure.

A review of history reveals that the power of steam had been recognized as early as 150 B.C., but it was left practically unharnessed until the 18th and 19th centuries.

At the present time, the majority of power in the world is produced by the use of the expansive energy in steam. Steam had been generated in closed vessels and used for the conversion of chemical to physical energy long before it was used for the generation of electricity. The development of the steam engine to drive line shafts in mills and pumps in the mines and for the propulsion of ships and locomotives went hand in hand with the developments and improvements in boilers.

Boilers were developed to do useful work by many inventors in various countries. Boilers in the 1700s and 1800s were most frequently built by inventors with little or no engineering training or education. For this reason, these boilers appeared in all different shapes: oval, rectangular, round, and combinations of these.

The materials for construction were limited to cast iron, iron, copper, and brass. The inventors of that early period were limited to these few materials, most of which were of poor quality and low strength. The development of boilers, therefore, depended on and kept pace with the improvements in materials.

The majority of boiler components were made of cast iron and iron that was produced in small foundries and forges. The terminology of today was unknown; it developed as the materials and boilers evolved. The processes used were often kept secret by the various producers.

The major metallurgical fuel in the United States in colonial times was charcoal. The iron thus produced was called charcoal iron, a term that lasted well into the 20th century. When one of the authors entered the boiler operating and inspection field in the 1940s, boiler tubes of charcoal iron were frequently specified for firetube boilers.

Due to the problems involved with transporting the material, iron works were located near a forest for production of charcoal and near a source of ore and limestone for the raw material itself. It was also necessary to have a source of power, so the works were most often located along a river or stream. One of the first of these plants was established in the Massachusetts Bay colonies in what was then Lynn, now Saugus. The Saugus Iron Works has been rebuilt and is now a part of the national park system. The plant was located on the banks of the Saugus River and consisted of water wheels, a furnace, bellows, and rolls. (See photos 1.1 to 1.3.) The materials were charcoal and bog ore and a gabbro rock for a flux. The air blast was supplied by bellows driven by a water wheel.

When the iron was ready, it was tapped by removing the clay plug in the bottom of the furnace. The iron was allowed to run into what amounted to a sand mold. The cast iron so produced was very brittle, so when strength and ductility was required, the cast bars were refined by reheating and hammering in the forge. The wrought iron bars were then ready for forming into a variety of products. Only narrow sheets, then called flats and rods, were produced in the slitting mills. These small mills sprang up throughout the colonies.

As engines were perfected and adapted to the locomotive, coal became more accessible and plentiful and started to be used in the furnaces to produce iron. The iron in these small blast furnaces was still refined by heating and pounding and forging. In the larger furnaces, production was limited to a few tons of iron a day. Production of plate was very difficult, and size was limited by the equipment and available power.

Restrictions placed on the colonies by Great Britain barred the development and installation of equipment such as rolling mills in this country. Although these laws were often ignored, they did retard the development of iron products. Most of the boiler plate manufactured in the 19th century was made of iron and produced in relatively narrow widths, generally less than four feet. The first charcoal iron boiler plate

Photo 1.1 The Saugus Iron Works. The furnace is at the left and the forge and rolls are housed in the building at the right. Photo by the authors.

Photo 1.2 The bellows and forge at the Saugus Iron Works. Photo by the authors.

Photo 1.3 The plate rolls at the Saugus Iron Works. Note the water-powered wooden drive gears. Photo by the authors.

rolled in this country was produced at the Lukens Rolling Mill prior to 1825.

Rolling mills were driven by water power until steam engines of sufficient horsepower were developed. In 1870, Lukens installed a steam-driven plate mill with a width between bearing supports of 84 inches. In 1891, the same company installed the first open-hearth plant.

As mentioned earlier, boilers were constructed in a bewildering variety of shapes. However, because of the frequency of failures, it eventually became evident that round boilers were the most reliable. Still, much controversy existed as to whether the tank boilers were better than the watertube boilers. The popularity of each type depended mainly on the use to which it was to be put.

Eventually the vertical firetube boiler came into use. Vertical boilers were also of many different sizes and designs. The vertical firetube boiler shown from a catalog illustration in photo 1.4 was built in sizes up to 60 inches OD, with maximum working pressures of up to 100 psi.

The Manning design for a vertical firetube boiler appeared around the 1890s (see sketch 1.1). It was designed with the firebox or furnace section several inches larger in diameter than the main section of the shell. The parts of the shell were joined by an Ogee ring. It was often claimed that these boilers produced steam that was superheated by as much as 25 degrees. (It is interesting to note that in the mid-1800s, some authorities took a stand against the use of superheated steam,

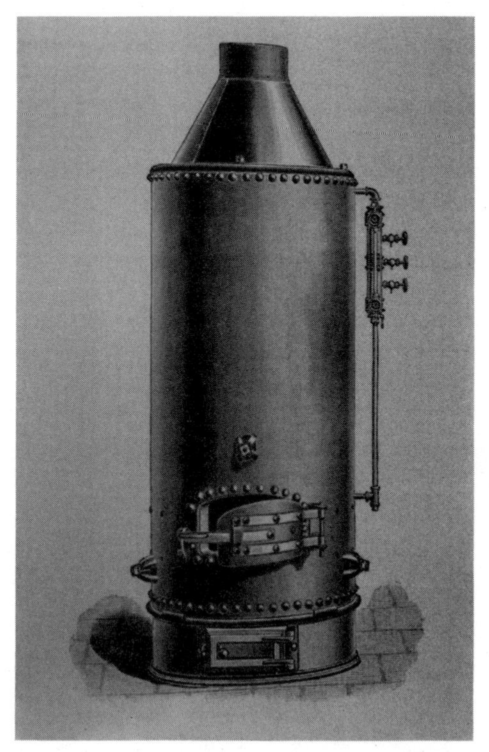

Photo 1.4 This shows a vertical firetube boiler, design circa 1903. The boiler was about 3 feet in diameter and 9 feet high. It had 144 square feet of heating surface and was rated at 10.3 horsepower. Photo courtesy of John Lynch, Hodge Boiler Works.

fearing that the steam would break down into hydrogen and oxygen and then explode. It was also reported by A. E. Seaton in his *A Manual of Marine Engineering* that the use of superheated steam was stopped because the pressure had gone above 60 psi. Myths such as this were greatly responsible for delaying the development of the use of superheated steam.) By 1900, these boilers were built for 125 psi, and unless the customer ordered the plates to be charcoal iron, they were made of steel. The tubes were lap-welded. Photo 1.5 shows four Manning boilers installed at the Amoskeag Manufacturing Company in Manchester, New Hampshire, around the turn of the century.

Improvements in the tank-type boilers by the addition of flues or internal furnaces and tubes greatly increased the efficiency. These became known as the horizontal return tubular internally fired return tubular boilers. (See photo 1.6.) They were the choice for marine service, but they were also widely used in stationary service. The Scotch

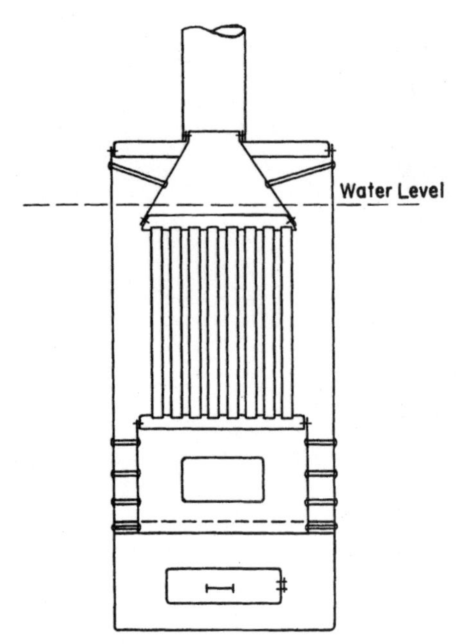

Sketch 1.1 Vertical tube submerged tube boiler. Courtesy Mutual Boiler & Machinery Insurance Company.

boiler, as it became known, consisted of a large-diameter shell with one or more circular furnaces extending from head to head (sketch 1.2). The hot gases passed from the furnace into the back connection. This area was often water-cooled ("wetback"), unlike the brick-back connection of the HRT boiler. The gases then reversed direction and passed to the front through tubes generally about 3 inches in diameter and then out to the smoke box and chimney. Because of the large diameters of the shells, pressures were limited by the materials for construction. Before the turn of the century, these boilers operated at 30 to 80 psi. The furnaces were subject to collapse due to the pressure on the external surfaces, so increases in pressure required them to be strengthened. Many methods were used that included various designs of corrugated furnaces such as the Morison, Purves, Leeds, Brown, and Fox. In the United States, the Morison corrugated furnace was the most commonly used. Another development was the Adamson ring, which permitted short sections of plain circular furnace, flanged at both ends, to be butted together with a stiffening ring between them. The sections were then riveted together to form the length of furnace desired. This design, now welded, is still specifically permitted in Section I.

Safety, ease of maintenance, and reliability were the main reasons for the popularity of the internally fired return tubular boilers. In

Photo 1.5 Four Manning design firetube boilers at the Amoskeag Manufacturing Company circa 1900. Courtesy John Lynch, Hodge Boiler Works.

Photo 1.6 Single furnace and double furnace, internally fired, horizontal return tubular (Scotch type) boilers, with corrugated furnaces. Courtesy John Lynch, Hodge Boiler Works.

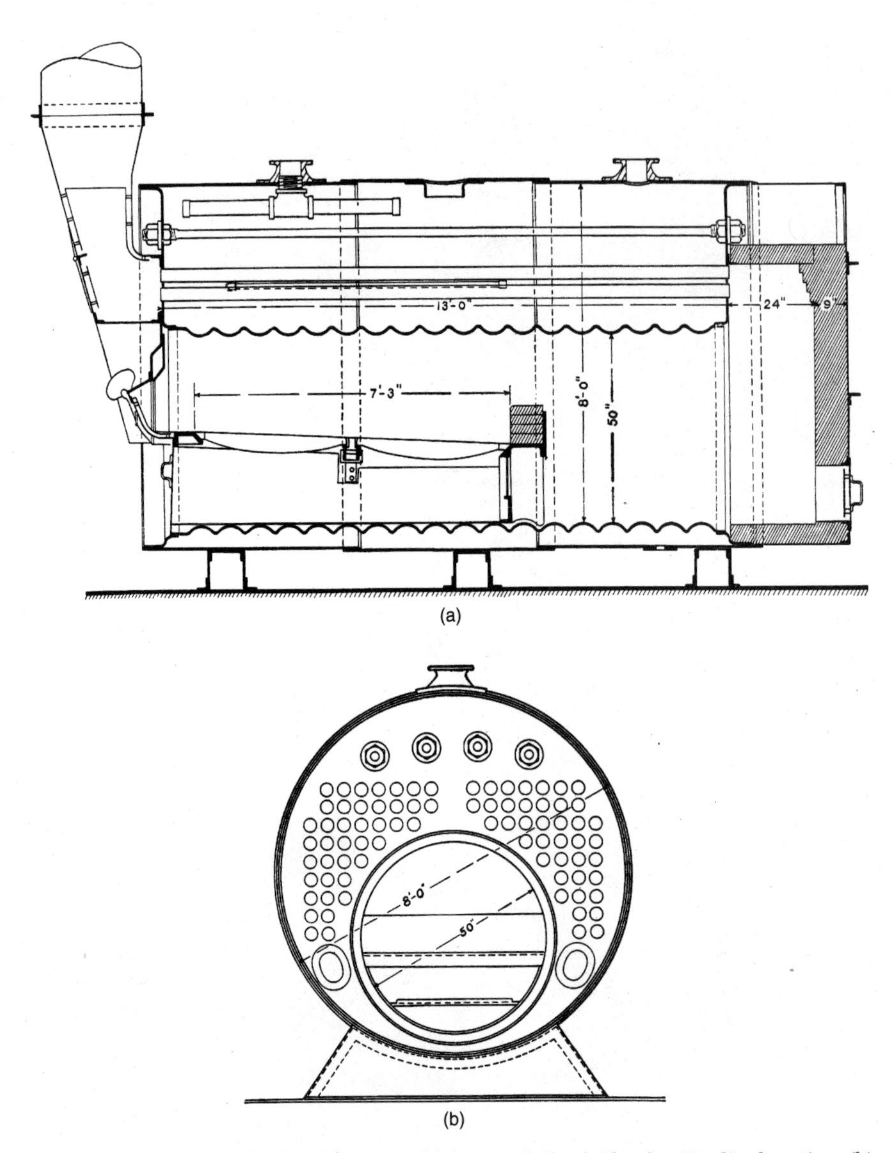

Sketch 1.2 (a) Internally fired horizontal return tubular boiler, longitudinal section. (b) Front elevation, breeching, and fire door removed. Courtesy John Lynch, Hodge Boiler Works.

fact, Scotch boilers are still the boilers of choice for building heating and anywhere steam is used for process and limited to about 150 psi or less. Most modern Scotch boilers have either plain circular furnaces or ring reinforced furnaces of the type described in PFT-17 (photo 1.7).

Photo 1.7 An externally fired horizontal return tubular boiler, the type known as HRT. Courtesy John Lynch, Hodge Boiler Works.

The HRT (horizontal return tube) boiler was also a marked improvement over the tank boiler in that the hot gases passed under the shell and then returned through tubes rolled into the front and rear heads. The gases then exited the boiler through breeching at the front to the smoke stack. This design was called the externally fired horizontal return tube boiler and became known as the HRT.

A 72-inch-diameter, 18-foot-long coal-fired HRT (sketch 1.3) would produce about 8000 pounds of steam per hour. Prior to the turn of the century, the shells and heads were made of iron. The shift to steel began to take place in the 1890s when steel of flange and firebox quality became more plentiful. Boilers of this vintage were rated in horsepower according to the ASME standard as "34.5 pounds of steam per hour from and at 212°F."

It should be pointed out that as boiler designs developed, methods of fabrication also had to be improved. Shells were made by rolling the material to the required diameter and then riveting the longitudinal joints. The shells were made in multiple courses, fitted together, and the circumferential seams single-riveted. The workmanship in achieving tightness, especially at the junctions, was of a high order.

The transportation industry was dependent on the development of the steam engine and the locomotive boiler (sketch 1.4). The locomotive boiler is a firetube boiler consisting of the barrel or shell and the firebox section. The furnace water legs are supported by stay bolts, as are the water legs of the vertical firetube boilers and the wetbacks of the Scotch boilers. The design of the locomotive boiler also makes use of stay bolts in the furnace, and crown bars and girder stays to support the top or crown of the furnace. One of the few remaining operating steam locomotives is at the Cog Railway ascending Mount Washington

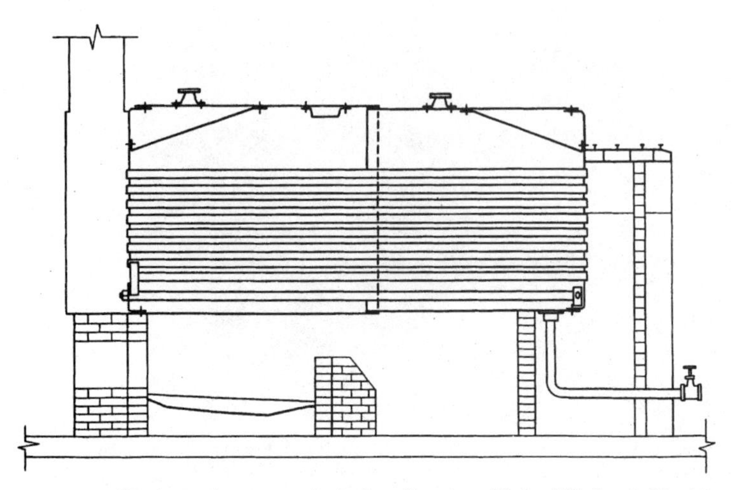

Sketch 1.3 Horizontal return tube boiler. Courtesy Mutual Boiler & Machinery Insurance Company.

in New Hampshire. At this writing, an order for two new locomotive boilers is being negotiated.

In the period lasting from approximately 1860 to well into the 1900s, the firetube boilers—the vertical firetube, the horizontal tubular boiler, and the internally fired return tubular boilers—gradually replaced most other types of boilers for plant use. However, a battle was waged by those favoring the watertube boilers. Many of the advancements in boilers of both types and in engines were being held back by the need for better materials.

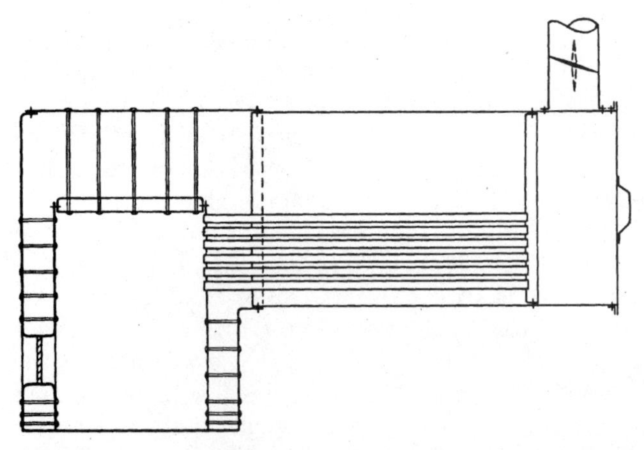

Sketch 1.4 Locomotive firebox boiler. Courtesy Mutual Boiler & Machinery Insurance Company.

The period from 1850 to 1900 saw many advancements in boiler design that took advantage of what materials were already available. Various configurations of watertube boilers were designed, incorporating cast iron steam drums and headers connected by copper or brass tubes. Early references call the tubes "lapped tubes," although the method of joining was actually forge welding. The tubes seldom lasted more than three to five years, and some references indicate failure was common in just a few months. Rarely were the causes of the failures indicated, but one can only believe many to have been failures of the lap welds and many others caused by overheating due to scale buildup.

William Blakey and John Stevens were among the first to investigate the generation of steam in boilers utilizing water tubes. In 1856, Steven Wilcox built a boiler resembling the firebox of a locomotive boiler, but with horizontal tubes running from front to rear. Ten years later George Babcock joined with Steven Wilcox, and together they patented the first Babcock & Wilcox boiler (sketch 1.5). This boiler consisted of horizontal cast iron tubes at the top—these served to collect the steam generated in inclined cast iron tubes. Later improvements included a horizontal drum of steel or wrought iron with cast iron chiseled heads, cast iron sectional headers, and steel or charcoal iron tubes. This boiler was developed in 1877.

Soon after the Bessemer process was developed in England in 1868, it was introduced in this country. This process had one major drawback in that it did not reduce the phosphorus content of the steel. Ore from most areas in America contained rather high percentages of phospho-

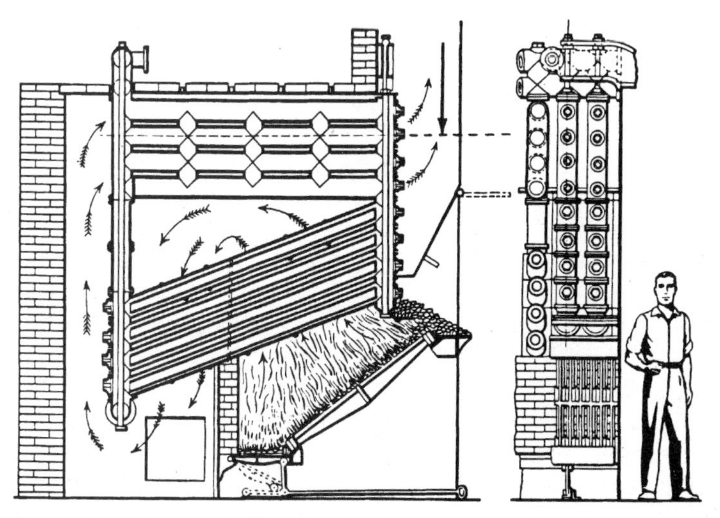

Sketch 1.5 Babcock & Wilcox's first boiler, patented in 1867. Courtesy Babcock & Wilcox.

rus, so the Bessemer steel produced was relatively brittle, especially when cold. In spite of this, the Bessemer process was the predominant method used until after the turn of the century. Shortly after the Bessemer process was introduced in America, the open hearth process followed. The open hearth process had the advantage of reducing the phosphorus content from the iron ore. This resulted in a greatly improved ductility of the steel.

Meanwhile, alongside Babcock and Wilcox, names well known to us today started to emerge: Stirling, Cahall, Hodge, Edgemore, Wickes, Cleaver Brooks, and many others.

The Cahall boiler was a vertical watertube boiler fired by coal in a Dutch oven. It consisted of two drums—one directly above the other—joined by straight tubes. This boiler, built by the Aultman and Taylor Machinery Corp. in Mansfield, Ohio, was well received in the 1890s. This design originally appeared in the March, 1897 book *"Cahall" Water Tube Steam Boiler.* At this time, there were no universal standards for materials or requirements for processing or production. Failures were the main factor in the search for more reliable materials and methods of design and construction of boilers.

Each supplier established his own standards and advertised his materials accordingly: Aultman and Taylor announced that their drums were made of the "best mild open hearth flange steel, connected by lap-welded tubes of the best charcoal iron." The quality of steel continued to improve, and boiler plate became known as flange and firebox plate. The term "firebox steel" came about because it was developed for use in the furnaces of locomotive boilers.

Realizing the need for other designs of boilers than the Cahall vertical tube boiler, Aultman and Taylor decided to explore other types of watertube boilers and decided the Babcock & Wilcox horizontal tube boiler was the best of the ones being produced. Aultman and Taylor adopted this design and produced it under the name Cahall Babcock and Wilcox. The shells and heads were again made of the "best open hearth flange steel." The edges were beveled, and the rivet holes were drilled after forming. The longitudinal joints were double-riveted for pressures up to 160 psi. For pressures from 160 to 250 psi, the longitudinal joints were double butt-strap, triple-riveted. The crossboxes attached to the drums had previously been either cast iron or formed from plate. Both processes had resulted in problems with cracking of the cast iron or the poorly fitted plate-formed crossboxes. On the Cahall Babcock and Wilcox boilers, the crossboxes were made by melting. Open hearth steel was melted and poured into molds and termed "flowed steel" (see figure). The crossboxes were annealed before machining. The sinuous headers for boilers operating at pressures above 225 psi were also made of "flowed steel." This is perhaps the earliest reference to the casting of steel boiler parts.

"FLOWED" STEEL.

It having come to our notice on several occasions recently that some of our competitors in the manufacture of water tube boilers have been offering to the general public Babcock & Wilcox type of boilers, the headers or manifolds and cross-boxes on which are claimed to be made of "flowed" steel, we desire to call your attention to the fact that "flowed" steel is a special mix of open hearth steel manufactured under a secret formula which belongs to us alone, the knowledge of the preparation of which is in the possession of no one except the Penn Steel Casting & Machine Co., of Chester, Pa., and as any one offering either headers, cross-boxes or flanges made of "flowed" steel is doing so with intent to deceive the public, we publish the foregoing statement and, in addition, the accompanying letter written to the Penn Steel Casting & Machine Co. by our general Eastern agents, and their answer to the same.

CAHALL SALES DEPT.

OFFICE OF THAYER & CO., INC., DREXEL BUILDING,
PHILADELPHIA, PA., *January* 21, 1897.

Penn Steel Casting & Machine Co.,
Chester, Pa.

GENTLEMEN: We have information that certain parties are claiming to be able to furnish—in fact, offer to furnish with their boilers—headers and other parts of "flowed" steel.

As "flowed" steel is of a special mixture and our property, we would ask if you have ever furnished or are now furnishing "flowed" steel headers to any party or parties other than ourselves?

An early reply will oblige, Yours truly,
(Signed) THAYER & CO., INC.

OFFICE OF PENN STEEL CASTING & MACHINE CO.,
CHESTER, PA., *January* 21, 1897.

Messrs. Thayer & Co., Inc.,
Drexel Bldg., Philadelphia.

GENTLEMEN: Referring to your inquiry of even date, we beg to state that we consider the special mixture for flowed steel we are making for you your property, of which you have the sole right, and that we never have nor never will furnish flowed steel to any one but yourselves, unless authorized by you. Neither will we give to any one any information as to the formula mixture of which this special mixture is made.

Yours respectfully,
PENN STEEL CASTING & MACHINE CO.
(Signed) FRED. BALDT, *Manager.*

"Flowed Steel" description from 1897 Cahall catalog.

Until the turn of the century, there was little emphasis in the United States on the uniformity and testing of materials or of the methods of determining the safe working pressures of boilers. In 1906, the need for a uniform boiler design, materials, and inspection code was dramatically brought to the attention of the public when two violent boiler explosions rocked the city of Brockton, Massachusetts. The destruction of the Evans Shoe Factory was followed by the Grover Shoe Company blast (photo 1.8). At the time of the explosion at the Grover Shoe Company, about 350 people were employed in the factory. With little or no warning, one of the horizontal return tubular boilers in the basement of the building violently exploded. Reports indicated that the boiler went straight up through the roof of the building, damaging the foundation and then breaking the building timbers in its path. After going up through the building, it went north on Denton Street. Ironically enough, the boiler went through the house of the chief engineer of the Grover Shoe Company before coming to rest. A total of four homes and six business buildings were destroyed, and others were damaged. Fifty-eight people were killed, and 50 were injured. The HRT boiler that caused all this damage was operating at 75 to 100 psi (see photo 1.8).

Based on currently available information, it does not appear as though an investigation was conducted to determine the exact cause of this explosion; however, the extent of the damage indicates that the boiler had a large quantity of water in it at the time (i.e., it was not a dry boiler overheating failure). It may be further concluded that the failure in the HRT shell itself was probably located at or near the bottom and caused by corrosion or local overheating because of scale buildup. Overfiring leading to excess pressure may have also been a cause. In any event, the need for a boiler design code, rules for periodic inspection of boilers, and provisions for the licensing of engineers and firemen had been made clear to the people of Massachusetts.

By an act of the Massachusetts legislature, the Steam Boiler Rules were formulated by the newly established Board of Boiler Rules. These rules were based on the best information available from all sources at the time, including boiler manufacturers, the U.S. Coast Guard, the U.S. Navy, Lloyd's Register of Shipping (Lloyd's Register of Shipping had for many years been involved in the inspection and classification of sailing ships. With the advent of the steam boilers and engines to propel ships, Lloyd's Rules had been expanded to include the methods of calculating, manufacturing, and inspecting steam boilers), and operating engineers and educators. The initial rules were put into effect immediately but required many revisions over the next few years. These rules set up requirements for annual internal and external inspections of all boilers coming under the scope of the law. Since this was a new procedure, boiler inspectors had to be hired and trained.

Photo 1.8 The Grover Shoe Company after the boiler explosion of 1905. Courtesy Mutual Boiler & Machinery Insurance Company.

Once the inspection process had been developed, the boiler inspectors were faced with the task of inspecting the boilers, measuring them, determining the methods of construction, and, as much as possible, identifying the materials of construction so that the maximum safe working pressures could be established.

Following the adoption of the Massachusetts Boiler Law in 1911, Ohio formulated its own code based on the Massachusetts one. Ohio was followed in 1918 by New York and then by many other states. The major problem that developed was that the rules varied from state to state and, in some cases, from municipality to municipality. This made it very difficult for the manufacturers of boilers and appurtenances to satisfy the conglomeration of different and sometimes conflicting rules and regulations.

The American Society of Mechanical Engineers started to establish methods for determining the safe working pressures on boilers and specifications for materials for use in boiler construction. Engineering colleges and universities such as the Massachusetts Institute of Technology provided much of the technical information and data necessary to establish the methods of calculating the safe working pressures of boilers and parts thereof. Much interest in the development of boilers was inspired in the American Society of Mechanical Engineers by George H. Babcock, a past president of the society. In 1911, a committee was formed to establish rules for the construction of boilers; these rules eventually became Section I, Power Boilers. The committee still exists as the Boiler and Pressure Vessel Committee. The Code that it writes and administers is generally recognized as the most comprehensive code for the design and fabrication of boilers and pressure vessels. The Code gradually became adopted by the majority of the states of the United States and the provinces of Canada. It is now the most widely recognized code worldwide.

The Babcock & Wilcox long drum boilers were also being produced. These drums were riveted, including the head to shell joints. Prior to the turn of the century, the longitudinal seams of boilers were lap-riveted.

The double butt-strap longitudinal joints started to be used in most high pressure watertube boilers shortly after 1900. Many explosions in the lap seam boilers built before 1903 were due to failures in the longitudinal joints. This phenomenon was rather thoroughly investigated and reported in the Mutual Boiler and Machinery Insurance Company publication entitled "The Causes and Prevention of Lap Cracks in Boilers" by H. M. Spring and D. E. Carroll. The production of vertical firetube HRTs and Scotch boilers continued to dominate the market for the smaller plants and buildings. It should be noted that throughout the first half of the 20th century, many of the larger office buildings and hotels operated their own power plants. (Many of these were phased out during World War II due to the shortage of fuel.)

A number of designs for water tube boilers emerged that incorporated longitudinal drums and straight tubes, some with sinuous headers and others with box headers. Most of these boilers operated at pressures of less than 200 to 250 psi.

The long drum boilers were continued in production into the twenties. The capacity of this type of boiler was increased by extending the number of tubes in the sinuous headers and the length of the tubes. It was also common to produce the long drum boilers with one, two, and three drums.

The Consolidated Boiler Company of Barberton, Ohio was purchased by Babcock & Wilcox in 1906. Consolidated manufactured the Stirling four-drum boiler, which Babcock & Wilcox continued to build (photo 1.9). Until 1900, boiler operating pressures were mostly under 100 psig, and because of difficulties with lubrication of the cylinders of steam engines, superheated steam was seldom used. With the advent of the steam turbine, the demand for higher pressures and superheated temperatures stimulated advances in boiler design. The Stirling design was a major improvement in watertube boiler design and may be said to have been the basis for the many variations of modern watertube boiler design.

As the capacities and operating pressures increased, it became necessary to develop techniques of treating boiler water to reduce the formation of scale and the corrosion of the tubes, drums, and headers. Pretreatment and internal water treatment methods had to be developed through extensive research.

Extensive research into the causes of caustic embrittlement by such researchers as W. C. Schroeder and A. A. Burke, U.S. Bureau of Mines; Dr. H. H. Uhlig of Massachusetts Institute of Technology; D. E. Carroll of the Mutual Boiler and Machinery Insurance Company; and many others developed methods of detecting conditions causing intercrystalline cracking by the use of the Schroeder embrittlement detector and inspection techniques to detect the presence of embrittlement cracking in existing boilers.

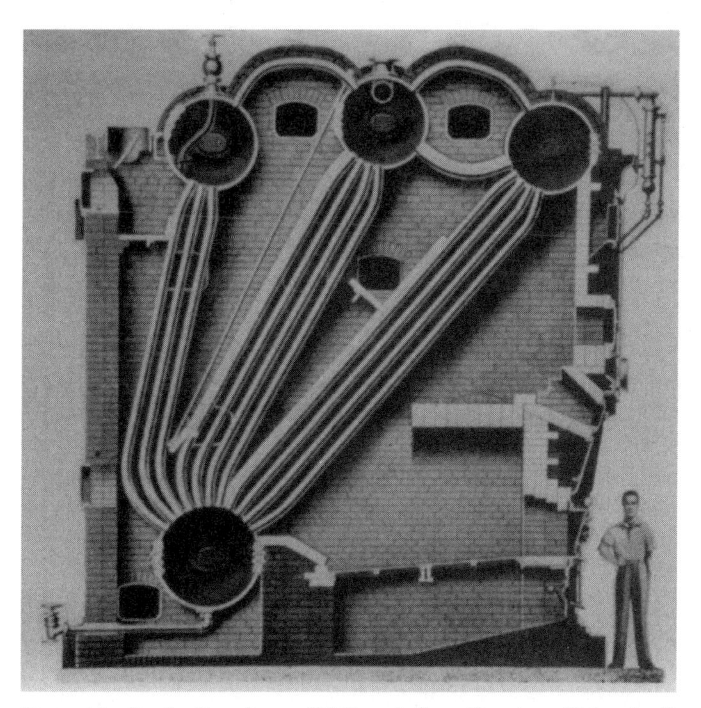

Photo 1.9 Early four-drum Stirling boiler. Courtesy Babcock & Wilcox.

During the earlier development of steam boilers, it was not uncommon for failures of tubes and other components resulting from corrosion and/or overheating to occur in as short a time as one to three or four years. With the high demand for reliable sources of steam for process industries and for power generation by the utility companies, methods of treating the boiler water became imperative. Not only chemical treatment but also the designs for deaerating and open heaters, evaporators, deionizers, and the like to reduce corrosion and scale formation were invented and refined.

Gradually the operating pressures in boilers increased to 400–600 psi until 1924 when Babcock & Wilcox designed and field-erected a boiler to have a maximum allowable working pressure of 1200 psi with a superheater and reheat superheater. This boiler was installed at the Edgar Station of The Boston Edison Company.

The development of boilers from the 1920s to the present time has taken many forms or combinations of steam drums, mud drums, headers, waterwalls, and tube configurations (see photo 1.10). Riveted construction was gradually phased out and welding took over as the method of choice. The rules for riveted construction were continued

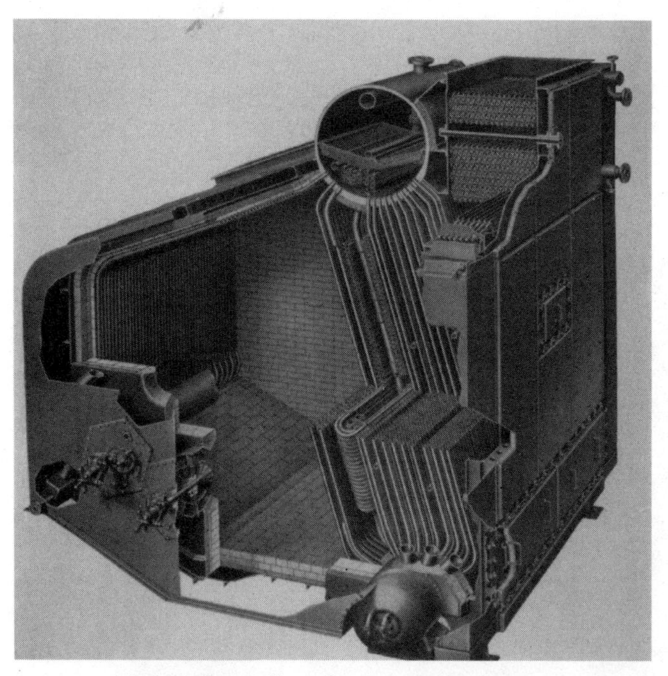

Photo 1.10 Relatively small modern watertube boiler, circa 1942. Courtesy ABB C-E Services.

in the Boiler Code through the 1971 edition. It should be noted that if a question regarding riveted construction comes up, the 1971 Power Boiler Code prevails today.

Although watertube boilers have come to dominate the market over the firetube boilers, there is still a place for firetube boilers. As of this writing, the Scotch boiler design variants still thrive for the smaller plants and heating applications. These designs were developed to fire the available fuels economically. Furnace configurations had to be designed to accommodate the various kinds of fuels: liquid, gaseous, and the many types of solids. The latter includes the different kinds of coal, each requiring special considerations in handling, burning, and disposing the ash and cleaning the boiler. A major advancement was

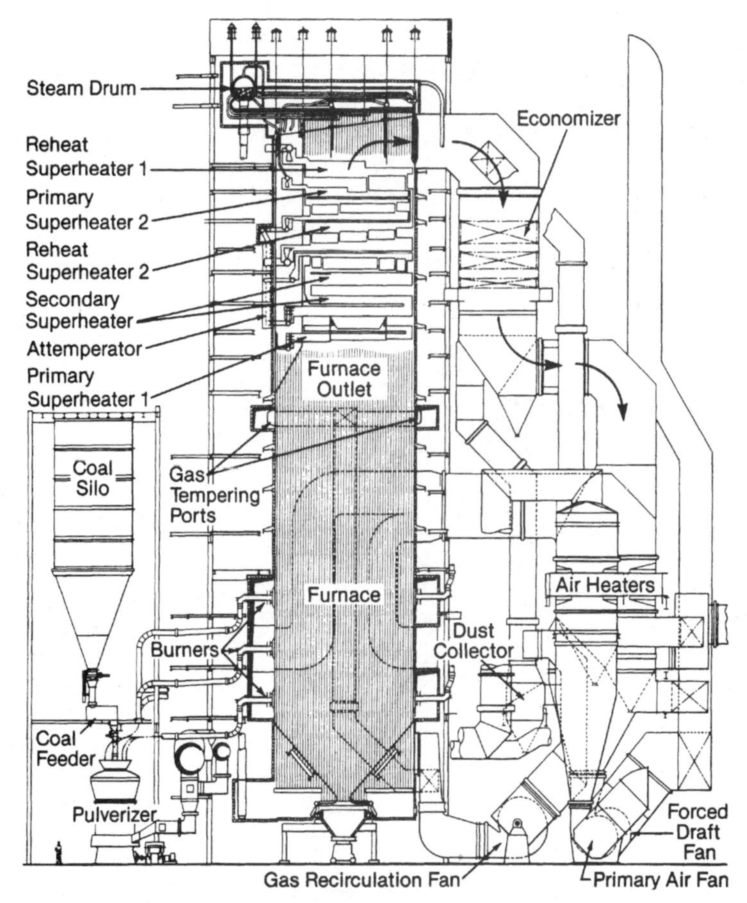

Sketch 1.6 A large modern watertube boiler for pressures of 1800 to 2400 psig. Courtesy Babcock & Wilcox.

the development of the black liquor boiler for the paper industry, which occurred in the late 40s or early 50s. As it turned out, these boilers are now called recovery boilers because a large percentage of the chemicals used in the pulp-making process are recovered and reused. The black liquor that had been discharged into the rivers is now being burned and the chemicals recovered for reuse.

By the second half of the 20th century, boiler technology had matured. Large power-generation boilers (see sketch 1.6) are generally drum-type watertube boilers, while small heating boilers are often Scotch-type firetube boilers. The rules for designing these boilers are long established, and the part of the Code concerned with boilers changes slowly. The new developments of the second half of this century have included nuclear power, environmental controls, and cogeneration. The nuclear code is Section III of the ASME Boiler & Pressure Vessel Code and now consists of 10 volumes. Environmental controls have been of immense importance and impact on the power generation field, but the effects on the rules for strength design of boilers, and therefore the subject matter of Section I and this book, has been minor or nil. Cogeneration is an attempt to improve power plant efficiency. The attainable efficiency of a heat engine (i.e., a boiler power plant) is limited by the second law of thermodynamics, and the limiting efficiency is dependent on the combustion temperatures. Boiler combus-

Photo 1.11 A modern shop-assembled watertube boiler. Note the steam drum at the top and the two headers at the bottom. Courtesy ABB-CE Services.

tion temperatures are limited by the materials available for use as boiler tubes, refractory, and the like. Higher combustion temperatures may be attained in devices such as gas turbines; the exhaust gases may then be conducted through boilers called waste heat boilers to extract more energy (see photo 1.11). Again, however, the impact of the new technology on the principles of design of boilers in accordance with Section I is quite minor. These relatively new areas of boiler technology are therefore outside the scope of Section I and this book.

As instructors of engineers and candidate boiler inspectors, the authors have become aware of the difficulties experienced by young engineers and engineers from foreign countries in understanding and applying the requirements of the ASME Code. Therefore, it is the intent of the following chapters to assist in answering the many questions that arise in the study and application of Section I.

Chapter

2

Cylindrical Parts Subjected to Internal Pressure

Pressure and Stress

Boilers are devices for generating steam under pressure for use external to themselves. In the Boiler Code, the English customary units still reign. The metric (SI) edition of the Code was discontinued with the 1986 edition. Pressure and stress are given in pounds force per square inch, usually abbreviated psi. In addition, it should be noted that the word pressure in a boiler means for most purposes the difference between the absolute internal pressure and the ambient or atmospheric pressure outside the vessel. This differential pressure is called the gage pressure, abbreviated *psig.*

Gage pressure is used in the design formulas in order to compute the required thickness of parts subjected to pressure. The numerical value used for the pressure in the design formulae is the MAWP or maximum allowable working pressure (PG-21). The term "design pressure" has the same meaning. It is important to note that boilers are not operated at their maximum allowable working pressures. One or more of the safety valves must be set at or below the MAWP (PG-67.3). Therefore, it is clear that the operating pressure must be less than the MAWP, or the safety valves would constantly be opening. The recommended differences between the operating pressure and design pressure (or MAWP) are not given in Section I. Guidance in this area may, however, be found in the National Board Inspection Guide in Appendix A, paragraphs A-202 and A-203, where recommended differentials for steam and high-temperature water boilers are given. In design practice, then, it is generally known what pressure is the desired operating pressure.

To this desired operating pressure is added the appropriate differential to obtain the MAWP. The value of the MAWP is then used to compute the required thicknesses of the various components of the boiler.

When a part is subjected to pressure, stresses are induced as the part resists deformation. The stresses will cause failure by excessive deformation or by fracture if they are too great. Therefore, for each material in the Code, there is an allowable stress. Designing so that the allowable stress is not exceeded provides reasonable assurance that the part will not fail, burst, or explode. The allowable stresses are generally lower at higher temperatures because metals lose strength at sufficiently high temperatures. The value of the allowable stress for a material is the lesser of: one-quarter of the lower value of the minimum specified tensile strength for that alloy at room temperature or the service temperature; or the lower value of two-thirds the minimum yield strength at room temperature or the service temperature. Thus, typically, a material having a room-temperature specified minimum tensile strength of 60,000 psi will have an allowable stress of 15,000 psi. This is equivalent to saying that the safety factor is 4, based on the ultimate tensile strength. It should also be clear that normally the bursting pressure for vessels designed by Code rules should be four times the MAWP, or somewhat higher if the materials have strengths higher than the specified minimum strengths.

At elevated temperatures, creep and stress rupture considerations govern the choice of allowable stress levels. Allowable stress becomes the lower of: the average stress to produce a creep rate of 0.01 percent in 1000 hours; or 67 percent of the average stress to cause rupture in 100,000 hours; or 80 percent of the minimum stress to cause rupture in 100,000 hours. A full discussion of the basis for establishing the stress values is given in Section II, Part D, Appendix 1.

The allowable stress values for code materials are found in Tables 1A and 1B in Section II, Part D (PG-23). The values in the tables are the values to be used in code design regardless of how the allowable stresses were determined.

Other sources of load on the pressure parts must also be considered in design. PG-22 indicates that the hydrostatic head must be considered (unless the particular formula specifically permits ignoring it) and that other loadings, such as those caused by the weight of the vessel and contents, must be considered if they increase the stress in the part by more than 10 percent of the allowable stress. Obviously, then, these other loadings have to be computed to determine whether they increase the stresses enough that they have to be "considered" by increasing the thickness to carry them.

Not all the materials for which stress values are given in Section II, Part D, may be used in Section I construction. Permissible materials for boiler plate are given in PG-6; for forgings in PG-7; for castings in PG-8; for pipes, tubes, and pressure parts in PG-9; and for stays in PG-13. Materials used in boilers must be made to one of the specifications listed in these paragraphs, except as permitted in PG-10, 11, or 12. (PG-10 describes the manner in which materials that were not produced to specifications listed in Section I may be shown, through documentation and testing, to be equivalent to listed materials. The exceptions in PG-11 and PG-12 are very narrow. PG-11 allows materials used in standard pressure parts to be either Section I materials or those listed in the American National Standard to which the part is manufactured (as long as the material is not specifically prohibited by Section I.) Standard pressure parts made to a manufacturer's standard must be made of Section I materials. PG-12 allows austenitic stainless steel for use in gage glass bodies. Materials selection for boilers, then, is normally made from PG-6, PG-7, PG-8, PG-9, and PG-13.

A great many pressure-containing parts are of cylindrical shape. The rules for designing cylindrical parts to carry internal pressure are given in PG-27. There are three types of calculations:

PG-27.2.1 gives the formulae for designing tubes. The tubes designed by these formulae may be no larger than 5 inches outside diameter and must also satisfy conditions given in the notes in paragraph PG-27.4, specifically Notes 2, 4, 8 and 10.

PG-27.2.2 gives the formulae for pipes, drums, and headers; that is, for all cylindrical parts that are not tubes, and for tubes greater than 5 inches outside diameter. The parts designed under this paragraph are also subject to Notes in PG-27.4: Notes 1, 3, 5, 6, 7, 8, and 9.

Both paragraphs are based on the thin wall assumption that a state of biaxial stress will exist in these vessels when they are subject to internal pressure. Formulae for thick parts, defined in PG-27.2.3 as having a thickness greater than one-half of the inside radius, are given in the Appendix, paragraph A-125. It should be noted that tubes are never thick in this sense.

In order to determine which of the two thin wall formulae should be selected to design a given part, one must be able to distinguish between tubes and pipes. Tubes are generally used as heat transfer devices, and they typically run from drum to drum, header to header, or tube sheet to tube sheet in boilers. They are made to tube specifications such as SA-178, SA-192, SA-209, SA-210, SA-213, SA-226, SA-250, SA-268, and SA-423. Tubes are generally produced to exact nominal diameters in order to fit tube holes. The thickness of tubes is often given as a Birmingham wire gage (bwg) number.

Pipes are generally used to transport fluids from one place to another. They are made to pipe specifications such as SA-53, SA-106, SA-335, SA-312, SA-369, or SA-376. Pipes are intended to be made up with threaded joints (but may be welded instead). In general the nominal diameter is neither the inside diameter (ID) nor the outside diameter (OD) for pipe. Pipe thicknesses are often expressed in terms of schedule numbers or "weight." For example, schedule 40 pipe is also known as standard weight. Pipes that are part of the boiler proper are designed by the Formulae in PG-27.2.2.

Boiler external piping is defined in the Preamble as the piping that begins where the boiler proper ends and extends through all Section-I-required valves. This piping requires Section I certification and inspection, but the design rules and materials are contained in ASME B31.1, Power Piping. The materials are ASTM materials, typically designated by A numbers rather than the SA numbers of Section II, Part A (i.e., A-53 rather than SA-53). The ASTM materials specifications are usually identical to the corresponding ASME specifications, but there are instances where the SA material has additional requirements beyond those for the corresponding A material. Sample problems for calculating tubes, pipe, and boiler external piping follow the derivation and discussion of the thin wall formulae.

Spheres

The stresses in a thin spherical shell may be calculated by recognizing that, by symmetry, all directions on a sphere are equivalent. The thin spherical shell under internal pressure can therefore be analyzed by cutting on any diameter and considering one resulting hemisphere as a free body (sketch 2.1). The variables are: P = gage pressure; R = inside radius; t = thickness; S = stress. The equation of force equilibrium is:

$$\pi R^2 P = 2\pi R t S$$

$$S = \frac{PR}{2t}$$

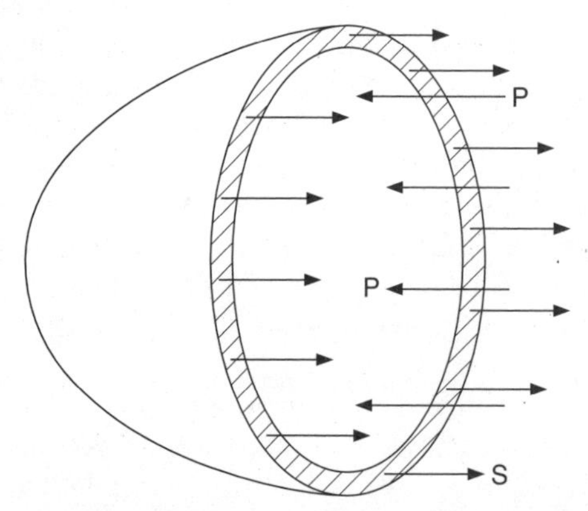

Sketch 2.1 Stresses in a thin spherical shell.

This is the classic formula for computing the membrane stresses in a spherical shell. It is based on the assumtion that the area of the shell carrying the stress can be considered equal to $2\pi Rt$. This formula was the actual Code formula for hemispherical heads as recently as 1949. As discussed in chapter 3, the current formulae for such heads are variants on the original.

Cylinders

The stresses in the longitudinal direction on a cylindrical vessel carrying internal pressure can readily be seen to be the same as those in a spherical vessel. Simply visualize the cylinder with a hemispherical head attached. These stresses are often considered as those tending to blow the heads off cylindrical vessels. The stresses in the circumferential direction may be found by isolating a half-cylinder of length L and writing the equation of force equilibrium (see sketch 2.2).

$$2PRL = 2StL$$

$$S = \frac{PR}{t}$$

Thus the stresses on a longitudinal seam are seen to be twice the stresses on a circumferential seam. In most cases, if one computes the cylinder on the basis of circumferential stresses, it is not necessary to consider the longitudinal stresses.

The tube formula PG-27.2.1 consists of three terms to find the thickness:

$$t = \frac{PD}{2S + P} + 0.005D + e$$

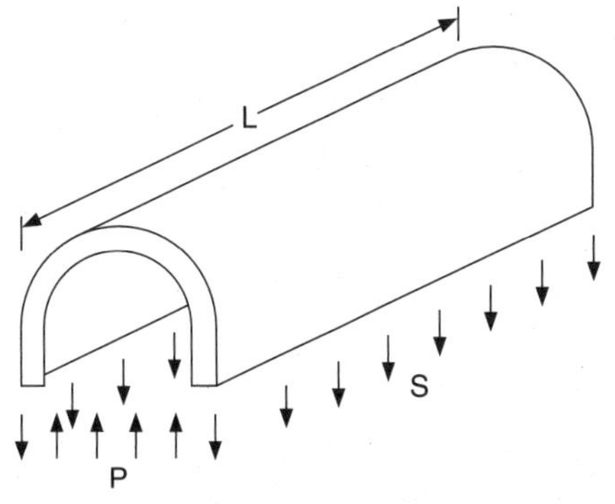

Sketch 2.2 Stresses in the longitudinal direction on a cylindrical vessel.

where D is tube outside diameter. The first term is almost the classic formula in terms of OD, $PD/(2S + 2P)$, but it is slightly altered in the conservative direction by reducing the denominator. The second term adds 0.005 inch of thickness to the tube for each inch of tube diameter. The third term, e, as seen below, adds additional thickness to tubes that are to be attached to a drum or tube sheet by expanding.

The pipe formula in PG-27.2.2, $t = PD/(2SE + 2yP) + C$, is also very close to the classic formula. In the denominator, the E term is 1 for pipe. The y term is temperature dependent but less than 1. It acts to increase the thickness slightly. The term C is an additive thickness for threading and/or strength, as shown in the sample problems.

Sample Problems

Sample problem 2.1. Thickness of a watertube (photo 2.1).

Find the minimum thickness of a watertube having 3-inch outside diameter if the MAWP is 400 psig. The material is SA-192. Consider two cases: (a) the tube is strength-welded to the tubesheet; (b) the tube is expanded and flared.

Solution: Use formula PG-27.2.1

$$t = PD/(2S + P) + 0.005D + e$$

where the terms are: t (thickness, inches); P (MAWP, given as 400 psig); S (allowable stress from Table 1A of Section II, Part D), and e (a term

Photo 2.1 View of the furnace of a Wickes watertube boiler. Photo by the authors.

adding extra thickness to tubes that are to be installed in the drum or header by expanding them into the tube seats).

To obtain numerical values for S and e, one reads the relevant Notes in PG-27.4. These are 2, 4, 8, and 10.

Note 2 indicates the temperature to be used in the table to find the value of S. The temperature should be not less than the mean metal temperature of the tube, but in no case less than 700°F if the tube absorbs heat. We therefore assume the mean metal temperature does not exceed 700°F and find the allowable stress in Part D of Section II, Table 1A, for SA-192 at 700°F, as 11,500 psi.

Note 4 defines the value of e, which is either 0 or 0.040 inch, depending on the tube diameter and thickness. For a 3-inch tube, the thickness is calculated using $e = 0$, and then, if t is less than 0.120 inch and the tube is to be expanded into the drum (rather than strength-welded to the drum), the extra 0.040 inch is added. If the tube is to be strength-welded to the drum, no extra thickness is required, and $e = 0$.

Note 8 allows rounding up to the next higher unit of 10 psi when computing the allowable pressure for a tube of definite minimum thickness (not applicable to this problem).

Note 10 permits the designer to ignore the hydrostatic head when computing the required thickness of tubes or when calculating the MAWP of tubes of known thickness. It is emphasized that this note does not apply to pipes or other cylindrical pressure parts calculated by the formulae in PG-27.2.2.

The solution is therefore:

$$t = \frac{(400)(3)}{(2)(11{,}500) + 400} + 0.005(3) + 0$$

$$= 0.066''$$

if the tube is strength-welded, as it is in case (a).

Then, per Note 4, 0.040 inch must be added if the tube is to be expanded into the drum, so the final answer for case (b) is $t = 0.106$ inch.

Alternatively, for tubes of SA-178 grade A, SA-192, or SA-226, if the mean wall temperature does not exceed 700°F, Table PWT-10 may be used to select the tube thickness for tubes expanded into the holes in the tubesheet (see PWT-10.3, Tube Wall Thickness). The thickness 0.095 inch is good to 390 psi. For 400 psi, however, the next higher thickness in the table—0.120—would be required.

The requirements for attaching the tubes to the tubesheet are given in PWT-11. If the tubes are to be expanded and flared, they must extend into the drum at least ¼ inch and not more than ¾ inch before flaring. After flaring, the outside diameter of the tube should be at least ⅛ inch larger than the tube hole. Expanding and seal welding may also be employed with or without flaring. The detail requirements for the various alternative combinations are given in PWT-11.1.

There are also mechanical requirements for preparing the tube holes. PG-36.9 and PG-79 require that the sharp edges of tube holes be removed on both sides of the plate (tubesheet) by filing or some other method. PG-79 also states the manner of making the tube holes. They may be drilled full-size in the plate. They may be punched at least ½ inch undersize and then drilled, reamed, or cut with a rotary cutter to the full diameter. If the holes are cut by means of a thermal process, they must be made sufficiently undersize that the machining to full-size removes all the material that had mechanical and/or metallurgical properties altered by the cutting.

Seamless pipe complying with SA-53 or SA-106 not over NPS 1½ inches can be used as tubing in watertube boilers, as described in PWT-9.2. If threaded tubes are used, PWT-11.3 requires that the threaded connection standards of PG-39.5 be complied with. These include the minimum number of threads for a connection, given as functions of pipe diameter and pressure in Table PG-39, and the maximum pressure limitations for different diameters of PG-39.5.2, which also states that threaded connections may not be used at temperatures above 925°F.

TABLE PWT-10
MAXIMUM ALLOWABLE WORKING PRESSURES FOR SEAMLESS STEEL AND ELECTRIC RESISTANCE WELDED STEEL TUBES OR NIPPLES FOR WATERTUBE BOILERS, WHERE EXPANDED INTO DRUMS OR HEADERS, FOR DIFFERENT DIAMETERS AND GAGES OF TUBES CONFORMING TO THE REQUIREMENTS OF SPECIFICATIONS SA-178 GRADE A, SA-192, AND SA-226

Wall Thickness, in.	Nearest Bwg. No.	Tube Outside Diameter, in.																	
		½	¾	1	1⅛	1¼	1½	1¾	2	2¼	2½	2¾	3	3¼	3½	3¾	4	4½	5
0.055	17 −	590	350	...	...	...	...	...	...	...	...	...	...	...	...	...	...	...	...
0.065	16	1090	670	470	410	350	...	...	...	...	...	...	...	...	...	...	...	...	...
0.075	15 +	1600	1000	720	620	550	430	...	...	...	...	...	...	...	...	...	...	...	...
0.085	14 +	...	1340	960	840	740	590	490	410	...	...	...	...	...	...	...	...	...	...
0.095	13	...	...	...	1990	1760	760	630	530	460	400	350	...	...	...	...	...	...	...
0.105	12 −	...	...	...	...	1980	1600	1340	1150	570	500	440	390	...	...	...	...	...	...
0.120	11	...	...	...	...	...	1870	1570	1340	1170	1040	930	840	460	420	390	...	...	...
0.135	10 +	...	...	...	...	...	...	1790	1540	1340	1190	1060	960	880	800	740	680	...	...
0.150	9 +	...	...	...	...	...	...	2020	1740	1520	1340	1200	1090	990	910	840	780	670	590
0.165	8	...	...	...	...	...	...	...	1940	1690	1500	1340	1210	1100	1020	940	870	760	670
0.180	7	...	...	...	...	...	...	...	...	1870	1660	1480	1340	1220	1120	1040	960	840	740
0.200	6 −	...	...	...	...	...	...	...	...	...	1870	1670	1520	1380	1270	1170	1090	950	840
0.220	5	...	...	...	...	...	...	...	...	...	...	1870	1690	1540	1420	1310	1210	1060	940
0.240	4 +	...	...	...	...	...	...	...	...	...	...	...	1870	1700	1570	1450	1340	1170	1040
0.260	3 +	...	...	...	...	...	...	...	...	...	...	...	...	1870	1720	1590	1470	1290	1140
0.280	2 −	...	...	...	...	...	...	...	...	...	...	...	...	2040	1870	1730	1600	1400	1240
0.300	...	...	...	...	...	...	...	...	...	...	...	...	...	...	2020	1870	1740	1520	1340
0.320	...	...	...	...	...	...	...	...	...	...	...	...	...	...	...	2010	1870	1630	1450
0.340	...	...	...	...	...	...	...	...	...	...	...	...	...	...	...	...	2000	1750	1550
0.360	...	...	...	...	...	...	...	...	...	...	...	...	...	...	...	...	...	1870	1660
0.380	...	...	...	...	...	...	...	...	...	...	...	...	...	...	...	...	...	1990	1760
0.400	...	...	...	...	...	...	...	...	...	...	...	...	...	...	...	...	...	...	1870
0.420	...	...	...	...	...	...	...	...	...	...	...	...	...	...	...	...	...	...	1980

GENERAL NOTES:
(a) These values have been calculated by the formula in PG-27.2.1, using allowable stress values at a temperature of 700°F, from Table 1A of Section II, Part D. Values above the solid line include an additional thickness of 0.04 in. to compensate for thinning of tube ends due to the expanding process.
(b) Where calculated allowable working pressures exceeded an even unit of 10 by more than 1, the next higher unit of 10 is given in the table.

The term "strength-welded" is not defined in Section I, and the required welding details for tubes strength-welded to the drum are not given. However, guidance may be found in Section VIII, Division I UW-20. Here, the terms "strength weld" and "seal weld" are defined and examples of satisfactory weld details are given. A strength weld is one in which the full axial strength of the tube can be carried by the weld. Leak tightness is also provided. A seal weld is one that is designed to provide only leak tightness, with the holding power being provided by expanding the tube into the hole. Figure UW-20 shows four acceptable methods for strength welding tubes to tubesheets. These are: (1) fillet weld only; (2) groove weld only; (3) equal groove and fillet; and (4) unequal groove and fillet. Formulae for determining the minimum groove and fillet sizes for the different cases are given in UW-20 (b) and (c).

Sample problem 2.2. Size of strength welds on tubes.

Find the minimum weld sizes for strength welding the tube of sample problem 2.1 to its drum. Consider two cases: (a) fillet weld only; (b) groove weld only.

Solution: (a) The tube is 3 inches OD. The thickness is taken as 0.075 inch, which is the next larger standard thickness than the calculated 0.066 inch minimum required thickness. Then the fillet weld leg size can be calculated by the formula from UW-20(c):

$$l = [0.56D^2 + 3.06t(D - t)f_w]^{1/2} - 0.75D$$

where D = tube OD = 3 inches; f_w = weld strength factor, the lesser of the values 1 or the the tube allowable stress divided by the tubesheet allowable stress, assumed to equal 1 for this problem; l = fillet weld leg size; and t = tube nominal thickness. Therefore:

$$l = [0.56(3)^2 + 3.06(0.075)(3 - 0.075)]^{1/2} - 0.75(3)$$

$$= 0.14'' \text{ (use } 3/16'' \text{ fillet weld)}$$

(b) Here, using the same data and symbols but inserting g (groove weld depth) for l, we get:

$$g = [0.56D^2 + 2.03t(D - t)f_w]^{1/2} - 0.75D$$

$$= [0.56(3)^2 + 2.03(0.075)(3 - 0.075)]^{1/2} - 0.75(3)$$

$$= 0.09'' \text{ (use } 1/8'' \text{ groove weld)}$$

We note that this result is consistent also with Figure PFT-12.1. (The groove weld should not be less than ⅛ inch or the thickness of the tube, whichever is greater. The fillet weld should be not less than ⅛ inch or the thickness of the tube, whichever is greater.)

Sample problem 2.3. Maximum allowable pressure on a watertube.

Find the MAWP on a 2.5-inch watertube with a thickness of 0.085 if the tube is SA-178, grade A material.

Solution: Use PG-27.2.1:

$$P = S\,\frac{2t - 0.01D - 2e}{D - (t - 0.005D - e)}$$

where P = MAWP; S = allowable stress; D = OD of tube = 2.5 inches; and e = 0 or 0.040 inch, per Note 4.

The Notes again are numbers 2, 4, 8, and 10. In accordance with Note 2 (in the absence of any contrary information), we assume that the mean tube wall temperature is 700°F, and find in Section II, Part D, Table 1A, S = 11,500 psi.

Note 4 indicates that if the tube is to be expanded (this is the normal method, so it would be assumed that the tube will be expanded unless there is contrary information), e = 0.040 inch for 2.5 inches diameter for thicknesses less than 0.105 inch.

Note 8 will permit the pressure calculated to be rounded up to the next unit of 10, and Note 10 permits ignoring the hydrostatic head in the calculation.

The result is:

$$P = 11{,}500 \times \frac{2(0.085) - 0.01(2.5) - 2(0.04)}{2.5 - [0.085 - 0.005(2.5) - 0.040]}$$

$$= 303 \text{ psig}$$

$$= 310 \text{ psig (per Note 8)}$$

We note that the answer to this problem could not have been obtained from Table PWT-10.

Sample problem 2.4. Section I pipe.

Find the minimum required thickness for a pipe, 6-inch NPS (nominal pipe size) to carry 500 psig if the pipe is to have (a) welded joints and (b) threaded joints, 8 threads per inch. The material is SA-53 grade B welded pipe, and the temperature does not exceed 650°F.

Much of the piping associated with boilers will be boiler external piping, but for all piping within the boiler proper, or for tubes in excess of

5 inches OD and for other cylindrical boiler parts, the formula of PG-27.2.2 shall be used to calculate thickness (see PWT-10.2).

Solution: This problem is therefore solved as a Section I piping problem:

$$t = \frac{PD}{2SE + 2yP} + C$$

where the terms are: t (minimum required thickness, inches); P (MAWP, psig = 500); D (outside diameter, inches. It must be noted that the Code doesn't give standard pipe dimensions, and 6-inch pipe is not 6 inches OD. The information must come from a handbook or from ANSI/ASME B36.1. The OD is 6.625 inches); S (allowable stress from Section II, Part D, Table 1A. S = 12,800 psi. This is actually the term SE—see below); E (efficiency of a welded longitudinal joint or ligament. Per Note 1 of PG-27.4, the efficiency of seamless or welded cylinders is 1.0. In fact, some welded pipe has a joint efficiency lower than 1.0, but this is taken care of by adjusting the values of the allowable stress in the Table. Thus the tabulated stress values incorporate a multiplication by an efficiency number where the efficiency of the pipe is less than 1.0. In this case, seamless pipe has an allowable stress of 15,000 psi. The efficiency of the welded pipe is 0.85, so the tabulated stress for the welded pipe is 12,800 or $0.85 \times 15{,}000$); y (a temperature coefficient, the value of which is obtained from PG-27.4, Note 6. For all materials, at temperatures of 900°F and below, $y = 0.4$); and C (an additive thickness beyond that needed to retain the pressure. The extra thickness may be a threading allowance or an allowance to provide adequate structural strength and stability. The value of C depends on the pipe diameter and whether the pipe is threaded or has plain ends. In this case, for 6-inch pipe with plain ends, $C = 0$. For threaded ends on 6-inch pipe, C = depth of thread, h; and $h = 0.8/n$, where n = number of threads per inch. Thus $C = h = 0.8/8 = 0.100$ inch).

The applicable Notes of PG-27.4 are numbers 1, 3, 5, 6, 7, 8, and 9. Notes 1, 3, and 6 have already been discussed.

Note 5 states that when steel pipe is threaded and used for steam pressures of 250 psig or more, the pipe must be seamless and no less than schedule 80.

Note 7 deals with the manufacturing tolerance given to pipe. Generally, the minimum thickness of pipe can be as much as 12.5 percent less than the nominal thickness. When ordering pipe, it is necessary to order a weight or schedule number great enough that 12.5 percent less than the nominal thickness is at least equal to the calculated minimum required thickness.

Note 8 allows rounding up to the next unit of 10 psig when calculating the pressure on a pipe, the minimum wall thickness of which is known.

Note 9 permits leaving backing strips in place on circumferential welds in cylindrical pressure parts (including pipe), but requires removal of backing strips on long seams.

The solution for part (a) is:

$$t = \frac{500(6.625)}{2(12{,}800) + 2(0.4)(500)} + 0$$

$$= 0.127''$$

Then, to specify the pipe, tolerances must be considered. 0.127/0.875 = 0.146 inch (Note 7). The weight or schedule ordered must have a nominal thickness of at least 0.146 inch. 6-inch schedule 40 (i.e., standard weight) pipe has a thickness of 0.280 inch and could be used.

To solve part (b), the computation is the same as the one from part (a), but the C term is additive: t = 0.127 inch + 0.1 inch = 0.227 inch. Then, per Note 7, 0.227/0.875 = 0.259. We observe that schedule 40 pipe is thick enough, but Note 5 requires that for threaded pipe for steam service over 250 psi, at least schedule 80 is required. Therefore, for 6-inch schedule 80, nominal thickness of 0.432 inch is the answer to part (b). We note that PG-39.5 Table PG-39 would require this pipe to have 10 threads engaged and to be attached to a minimum plate thickness of 1.25 inches. We also note that 6-inch threaded pipe could not be used to attach boiler external piping (PG-39.5.2).

Sample problem 2.5. Boiler external piping.

A blowoff line for a boiler having MAWP = 1200 psig is 2-inch schedule 80. The hydrostatic head is 14 psi. The material is SA 106, grade B, and the pipe has plain ends. The corrosion allowance is 0.060 inch. Find the design pressure that should be used to calculate the required thickness. Does the given pipe meet the code requirements?

Solution: Design requirements for boiler external piping are given in paragraph 122.1 of B31.1. The topics include steam piping, feedwater piping, blowoff and blowdown piping, boiler drains, and miscellaneous boiler external piping. Blowoff piping is covered in 122.1.4 A.1: "The value of P to be used in the formulas in paragraph 104 shall exceed the MAWP of the boiler by either 25 percent or 225 psi, whichever is less, but shall not be less than 100 psig." Paragraph 101.2.2 requires that static head be included in the design pressure. Therefore the design pressure for this pipe is the smaller of 1200(1.25) + 14 or 1200 + 225 + 14. P = 1439 psig.

For the 2-inch NPS 80 pipe, the outside diameter is 2.375 inches, and the nominal thickness is 0.218 inch. The design formulae for straight pipe are given in paragraph 104.1. In this case we can calculate the required thickness and compare it to the thickness of the given pipe, or we can use the given pipe to compute the maximum allowable design pressure and compare the two to see if the pressure computed equals or exceeds 1439 psi. In either approach we must employ A.1.1: "If the pipe is ordered by its nominal wall thickness, the manufacturing tolerance on wall thickness must be taken into account." From Section II, Part A, or from ASTM A-106, we find that the manufacturing tolerance on the thickness of the pipe is 12.5 percent. Therefore we must reduce the nominal thickness by 12.5 percent for purposes of calculation.

So we have:

$$tm = \frac{PDo}{2(SE + Py)} + A$$

where tm = required minimum thickness; P = design pressure = 1439 psig; Do = outside diameter = 2.375 inches; SE = allowable stress on A106 grade B pipe at the design temperature (paragraph 101.3.2 states that the design temperature shall be assumed to be the same as the fluid temperature unless tests support a different value—therefore the design temperature here is the saturated steam temperature at 1439 psig or 1454 psia, 594°F; so from Table A-1, the allowable stress for A106 grade B at 600°F = 15,000 psi); y = temperature coefficient from table 104.1.2(A) = 0.4; and A = additive thickness term to compensate for material removed in threading, to provide for mechanical strength, or for corrosion allowance. In this case A = 0.060 inch. So:

$$\begin{aligned} tm &= \frac{1439(2.375)}{2[15{,}000 + 0.4(1439)]} + 0.060 \\ &= 0.170'' \end{aligned}$$

The actual minimum thickness is 0.191 inch, so the pipe does meet the requirements.

Alternatively,

$$\begin{aligned} P &= \frac{2SE(tm - A)}{Do - 2y(tm - A)} \\ &= \frac{2(15{,}000)(0.191 - 0.060)}{2.375 - 2(0.4)(0.191 - 0.060)} \\ &= 1731 \text{ psi} \end{aligned}$$

Since this is greater than 1439, the pipe does meet the requirements of the Code.

We note, then, that for boiler external piping, the design formulae, the value to be used for the design pressure for the different systems, and the allowable stresses for the materials at temperature are all found in B-31.1.

Other cylindrical pressure parts besides tubes and pipes are designed in accordance with PG-27. These include parts known as shells and drums. Examples might include calculating the minimum thickness for the shell of an electric boiler, where the heat is supplied by immersion heaters, or finding the thickness of the steam drum in a watertube boiler. Examples follow.

Sample problem 2.6. Thickness of boiler shell.

Find the required minimum thickness for the shell of an electric boiler if the inside diameter is to be 16 inches and the MAWP is 100 psig. The material is SA-285, grade C. The boiler has one longitudinal welded seam and two circumferential head-to-shell seams.

Solution: Use PG-27.2.2:

$$t = \frac{PR}{SE - (1 - y)P}$$

where t = required thickness; P = MAWP = 100 psig; R = inside radius = 8 inches; S = allowable stress ($T < 650°F$) = 13,800 psi; E = 1 per PG-27.4 Note 1; and y = 0.4 per PG-27.4 Note 6. Note that the C term applies to pipe but not to this cylindrical shell.

$$t = \frac{100(8)}{13{,}800 - (1 - 0.4)(100)}$$

$$= 0.058''$$

Then, per PEB-5.2, the minimum thickness of steel plates subject to pressure in electric boilers is 3/16 inch.

Ligaments and efficiency

In watertube boilers there is commonly at least one steam drum at the top of the boiler and a mud drum at the bottom. Boiler tubes run between these drums, so the drums must have holes in them for the tubes to enter. These holes are generally 1/64 inch to 1/32 inch larger than the OD of the tubes to provide clearance for insertion. The tube holes are large enough and numerous enough that they affect the strength of the cylindrical drum. The holes are in a definite pattern and the strength reduction they cause is compensated by increasing the thickness of the perforated part of the drum, commonly called the tubesheet, through which the tube ends pass. The metal between the tube holes is called the ligament, and the ratio of the strength of the perforated drum to an unperforated drum of the same thickness is called the ligament efficiency. The thickness of the tubesheet is then calculated using PG-27.2.2:

$$t = \frac{PD}{2SE + 2yP}$$

with the ligament efficiency used as the numerical value for E. If the drum is a steam drum, the tubes will all enter the bottom half, and there is no need for the unperforated top half of the drum to be increased in thickness. The thickness of the unperforated half of a drum, commonly called the shell, is calculated using PG-27.2.2, with $E = 1$.

The shell and tubesheet are made from plate rolled to a semicircle. They are joined to form a cylinder by two longitudinal welds, as shown in sketch 2.3. PW-9.3 states that the centerlines of the plates must line up, and the thickness transition must have a 3-to-1 taper, inside and out (see Figure PW-9.1 and sketch 2.3). PG-80.1 states that the maximum out-of-roundness for a cylindrical shell exposed to internal pressure shall be 1 percent of the mean diameter.

The ligament efficiency is calculated by the rules in PG-52. PG-52.2.1 gives the formula for the efficiency when the longitudinal pitch is the same for all tubes:

$$E = \frac{p - d}{p}$$

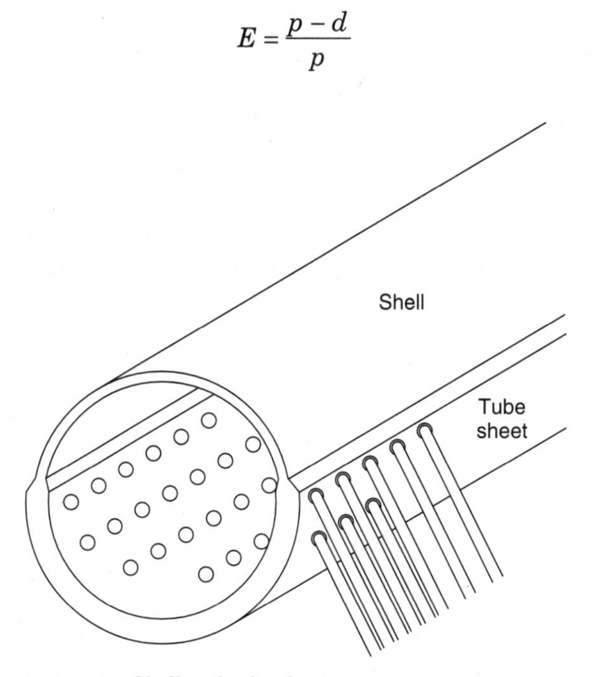

Sketch 2.3 Shell and tube sheet.

where p is the pitch of the tubes and d is the tube hole diameter. As shown in sketch 2.4, this is simply the ratio of the metal area with the tube holes drilled to the area without drilling holes. The undrilled area for length p is pt. After drilling, one tube hole diameter of metal is gone, so the remaining area is $(p - d)t$. The area ratio is thus: $(p - d)t/pt$, or $(p - d)/p$.

When the pitch is not the same for adjacent tubes but there is a repeating pattern, the code refers to the pitch of the pattern as p'. That is, p' is the length of metal in one pattern repetition. Therefore, the available metal in the undrilled state is $p't$. Some number of tube holes, n, are drilled in the pattern, so the area removed is nd. The efficiency is then given as $E = (p' - nd)t/p't$, or $E = p' - nd/p'$ (PG-52.2.2). Note that this has the effect of assuming that the holes are all drilled at the same pitch, the average pitch within the pattern (see sketch 2.5).

The foregoing explains the manner in which the efficiency E is calculated for use in the formula PG-27.2.2 to find the required thickness of a cylinder with a pattern of holes along its length. There will generally be more than one such row of tubes and tube holes, so the circumferential spacing of the rows must also be considered. PG-52.3 states that the strength of those ligaments subjected to a longitudinal stress shall be at least one half the required strength of those ligaments subjected to a circumferential stress. This is because the circumferential stresses are twice the longitudinal stresses in a thin cylindrical vessel. To meet this requirement, the metal area between the tubes in the circumferential direction must be at least one half the (average) metal area between the tubes in the longitudinal direction. This condition will be met if the circumferential pitch is at least one-half the average longitudinal pitch plus one-half the tube hole diameter ($p_c = (p_1 + d)/2$). It should be noted that the common practice of making the circumferential efficiency 50 percent of the longitudinal efficiency does not provide half as much metal area in the circumferential ligaments as in the longitudinal ligaments (see sketch 2.6). PG-52.3 will be met, however,

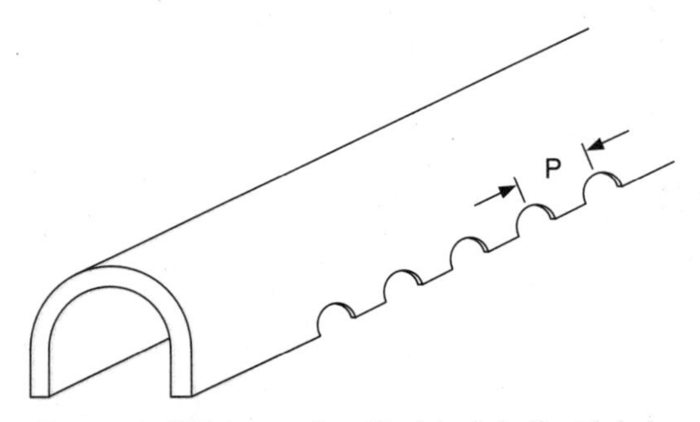

Sketch 2.4 Efficiency of a cylindrical shell with holes in a pattern.

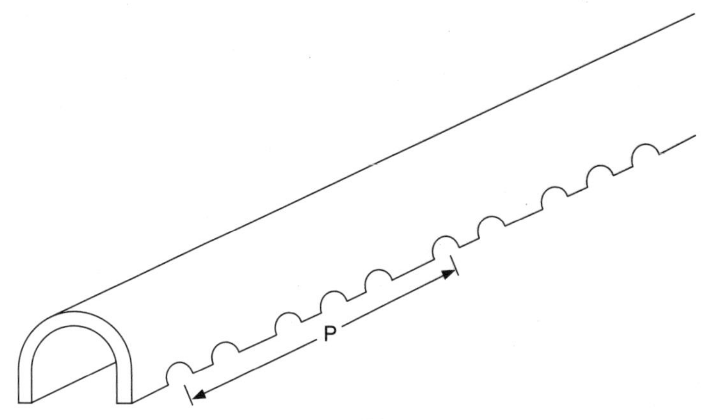

Sketch 2.5 Efficiency with varying pitch in a pattern.

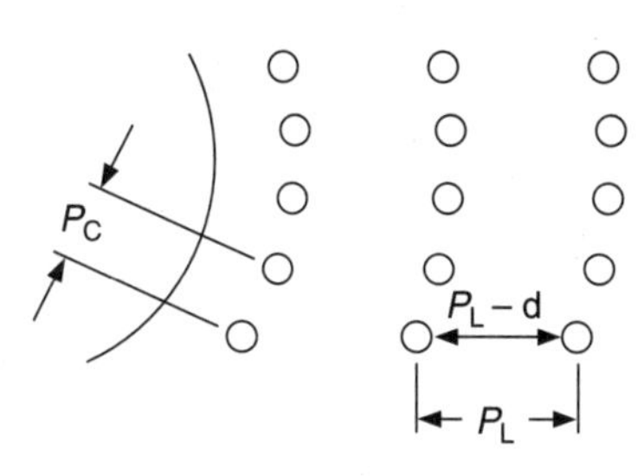

Sketch 2.6 Circular pitch and PG-52.3. PG-52.3 requires that the circumferential ligament be at least one half the longitudinal ligament. Therefore, the circular ligament = $(P_L + d)/2$. $P_c = [(P_L - d)/2] + d = (P_L + d)/2$. Then: $E_C = [(P_L + d)/2 - d]/[(P_L + d)/2] = (P_L - d)/(P_L + d)$. $E_C = [P_L/(P_L + d)] \times E_L$ for equal stresses in the ligaments. When P_L is not the same for all tubes, use the average longitudinal pitch $(P_L - nd)/n$.

if the circumferential efficiency is at least equal to the longitudinal efficiency multiplied by the longitudinal pitch divided by the sum of the longitudinal pitch plus the tube hole diameter. $E_c = E_1(p_1)/(p_1 + d)$.

Sample problem 2.7. MAWP on a watertube boiler drum (photo 2.2).

Find the MAWP on a watertube boiler drum if the shell thickness is ⅞ inch, the tubesheet thickness is 1⅜ inches, the tube holes are 3⅟₃₂ inches diameter and spaced alternately 5 inches and 6¾ inches. The material is SA-515 grade 70, and the inside radius of the tubesheet is 20 inches. Find the minimum circumferential pitch.

Solution: The allowable pressure for the shell and tubesheet must be computed, and the smaller of the two will be the allowable pressure on the drum.

For the shell, use PG-27.2.2:

$$P = \frac{SEt}{R + (1 - y)(t)}$$

Photo 2.2 View into a steam drum. Photo by the authors.

where P = MAWP; S = allowable stress from Section II, Part D (for SA-515, grade 70 at $T < 650°F$, S = 17,500 psi), E = ligament efficiency (PG-27.4, Note 1); t = plate thickness (⅞ inch and 1⅜ inch); R = inside radius of plate (R = 18 inches for the tubesheet and R = 18.5 inches for the shell per PW-9.3); and y = coefficient from PG-27.4, Note 6 = 0.4. Furthermore, C does not apply to shells per Note 3, and Notes 5, 7, and 8 do not apply here.

Therefore, for the shell:

$$P = \frac{(17{,}500)(1)(0.875)}{18.5 + (1 - 0.4)(0.875)}$$

$$= 805 \text{ psi}$$

For the longitudinal efficiency, use PG-52.2.2:

$$E = \frac{p' - nd}{p'}$$

where p' = 11.75 inches and n = 2 (see sketch 2.5):

$$E = \frac{11.75 - 2(3.031)}{11.75}$$

$$= 0.484, \text{ or } 48.4\%$$

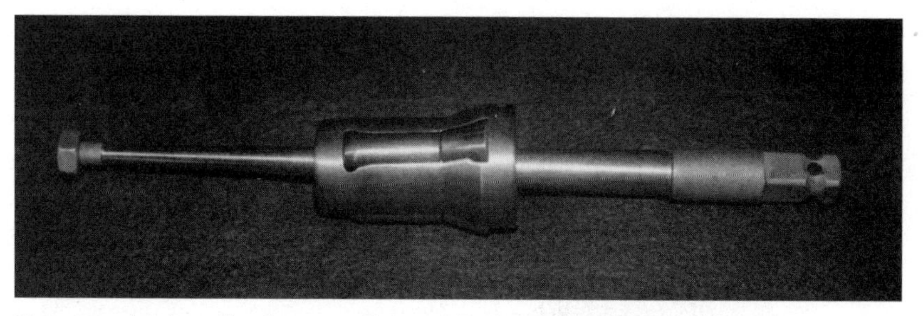

Photo 2.3 A tube roller for expanding and flaring tubes. Photo by the authors.

For the tubesheet:

$$P = \frac{(17{,}500)(0.484)(1.375)}{18 + (1 - 0.4)(1.375)}$$

$$= 619 \text{ psi (the lower value)}$$

The circumferential pitch of the tubes will be uniform. Setting the circumferential efficiency at half the longitudinal efficiency will give:

$$\frac{pc - d}{pc} = \frac{E_l}{2}$$

$$\frac{pc - 3.031}{pc} = 0.242$$

$$pc = 3.999''$$

The average longitudinal ligament is [11.75 – 2(3.031)]/2 or 2.844. Therefore setting the circumferential ligament at one-half the average longitudinal ligament means that the circumferential ligament must be at least 1.422 inches, which makes the circumferential pitch 1.422 + 3.031 or 4.453 inches. Note that setting the circumferential efficiency at one-half the longitudinal efficiency does not satisfy PG-52.3. Rather, to have the circumferential ligaments equal to one-half the (average) longitudinal ligaments in this case, the circumferential efficiency must be 0.319.

The third of the equations for designing cylindrical components subject to external pressure is for use when the thin wall assumption used in deriving PG-27.2.1 and PG-27.2.2 is not valid. PG-27.2.3 states that when the thickness exceeds one-half the inside radius, cylindrical parts shall be designed in accordance with the formula given in A-125 in the Appendix. (This formula is used for temperatures up to the critical temperature for steam, 705.4°F.)

Sample problem 2.8. Thick cylinder.

A header is 10 inches OD and 2 inches thick. The material is SA-266 Grade 1, and the maximum temperature is 650°F. Find the MAWP.

Solution: Note that the inside radius is 3 inches and the thickness is greater than half this measure. Therefore per PG-27.2.3, A-125 gives:

$$P = \frac{SE(Z-1)}{Z+1}$$

where P = MAWP; S = allowable stress from Section II, Part D (S = 15,000 psi); E = efficiency = 1.0; and $Z = (R_o/R)^2 = (5/3)^2 = 2.778$.

Substituting, we get:

$$P = \frac{(15{,}000)(2.778-1)}{2.778+1}$$

$$= 7059 \text{ psig}$$

Chapter

3

Heads

Heads most commonly form the end closures on cylindrical vessels. Therefore most heads are circular in contour, but other shapes are also commonly encountered. Section I contains the design criteria for the four most frequently used types of heads.

There are three types of dished or formed heads, usually used with pressure on the concave side. The reason for using formed heads is to take advantage of the inherently greater resistance to internal pressure of spherical or approximately spherical shapes. These dished heads are designed in accordance with the rules in PG-29. The shapes are shown in sketch 3.1.

Flat heads (and covers) may be circular, square, obround, elliptical, rectangular, or otherwise noncircular in shape. Flat heads are designed by the rules given in PG-31.

The most commonly used formed head is the dished head, also called a standard dished head, a standard head, or sometimes an ASME head. The shape is torispherical, with a central portion dished to a segment of a sphere and a knuckle portion at the edge of the central portion. The knuckle has a radius that is also called the corner radius in PG-29.13. A further requirement states that there must be an integrally formed cylindrical section (called a flange) after the knuckle. The required thickness or the maximum allowable working pressure for a dished head is calculated by the formula in PG-29.1:

$$t = \frac{5PL}{4.8S}$$

where t is the minimum thickness in inches of the head; P is the MAWP, psig; L is the inside radius in inches to which the head is dished; and S is the allowable stress at temperature, found in Section

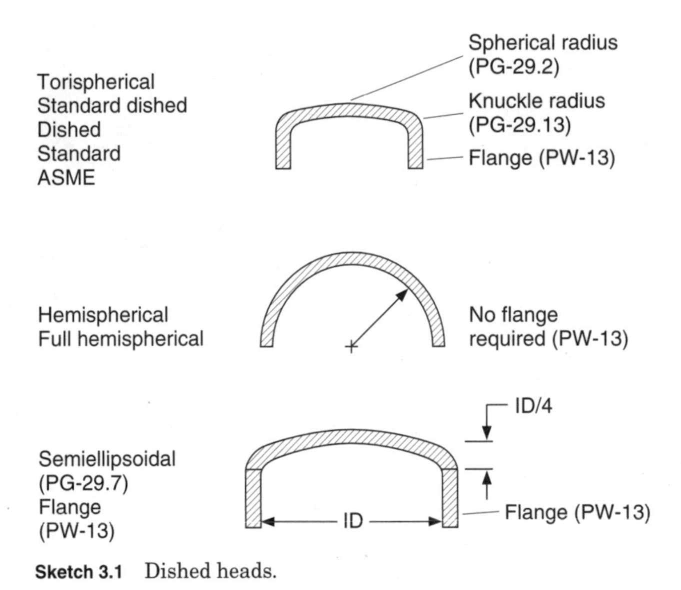

Sketch 3.1 Dished heads.

II, Part D, table 1A. (We repeat our caution from chapter 2 regarding the use of the tables in Section II, Part D.)

Dimensional and Proportional Requirements for Standard Heads

1. PG-29.2 states that the inside radius of dish of the spherical portion shall not be greater than the outside diameter of the flanged portion of the head. (This requirement prevents the head from becoming too close to a flat head.) Sometimes a head may be formed with a central section dished to one radius and the balance of the surface formed to a different radius. Where this construction is used, the thickness must be computed using the larger of the two radii. This results in a larger thickness or smaller MAWP.

2. PG-29.13 states that the (inside) corner radius (knuckle radius) must be at least equal to the greater of three times the thickness or 6 percent of the diameter of the shell. (The Code in some places rigorously specifies inside or outside diameter and in other places is rather vague. In PG-29, the expression "diameter of the shell" may be taken to mean the outside diameter of the flanged portion of the head.) It is extremely important that the minimum required knuckle radius not be reduced to minimize the chances of fatigue cracking. This type of cracking was common in heads with short knuckle radii in older-type boilers such as the early Stirling.

In the forming process, some thinning of the knuckle areas of heads can take place. PG-29.13 states that the thinning of the knuckle may not exceed 10 percent of the calculated required thickness (i.e., the thickness may be no less than 90 percent of the calculated value in the knuckle area). PG-29.10 requires that when the flange is machined to fit up with the shell, the thickness may not be reduced below 90 percent of the thickness calculated for a blank head.

3. PG-29.5 states that if a flanged-in manhole or other flanged-in opening over 6 inches is to be installed in the head, the value of the term *L* (the radius of dish of the spherical portion of the head) in the formula in PG-29.1 must be at least 0.8 times the shell diameter. We note that a flanged-in manhole may be installed in a dished head having a dish radius as large as the outside diameter of the flanged portion of the head. In this case, the value of *L* in the formula would be the actual dish radius. In no case can the value of *L* be less than the actual value of the radius of the dish; PG-29.5 makes the additional requirement that this value never be less than $0.8D$, even if the radius of the dish is less, for a dished head with a flanged-in manhole. The common error of simply setting the value of *L* at $0.8D$ can result in heads that are thinner than Code requirements for the rated MAWP. This can be seen by observing that the required thickness in the formula of PG-29.1 is directly proportional to the term *L*.

Access to the interiors of boiler drums is needed during both initial assembly and in-service inspection, maintenance, and repair. Flanging in is a very common way of installing manholes (and other openings) in heads. A flanged-in manhole is made by forming the edges of the opening in the head inward, as shown in photo 3.1. The proportions of the flanged-in opening are given in PG-34.2. The opening must be flanged to a depth of not less than three times the thickness of the head for thicknesses up to 1½ inches. For thicknesses over 1½ inches, the depth of flange must be at least 3 inches plus the head thickness. The depth of the flange is measured with a straightedge across the opening, as shown in sketch 3.2.

4. PG-44.1 states that the minimum size of an elliptical manhole shall be 12 inches by 16 inches. The minimum size of a circular manhole is 15-inch diameter. (The minimum size for an elliptical opening was formerly 11 inches by 15 inches.)

5. PG-44.3 states that the minimum width of the gasket-bearing surface on a manhole opening shall be 11⁄16 inch, and the maximum compressed gasket thickness shall be ¼ inch.

6. PW-13 states that dished heads (other than hemispherical heads, which are discussed in the next section) must have a length of flange of at least 1 inch for outside diameters up to 24 inches and at least 1½ inches for diameters over 24 inches. (We again note that the flange here

Photo 3.1 Steam drum with flanged-in manhole. Photo by the authors.

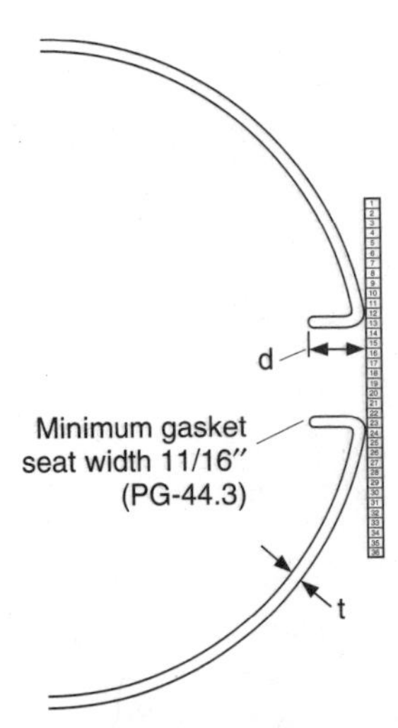

Sketch 3.2 Use of a straightedge to check flange depth of a flanged-in manhole.

is the cylindrical extension of the formed head, made to facilitate welding to the cylindrical shell. This is not to be confused with the flange depth in a flanged-in opening.)

7. PG-29.3 states that when a flanged-in opening that is over 6 inches in any dimension is installed in a standard head, the thickness of the head is to be calculated in accordance with the formula in PG-29.1 and then increased by the larger of $\frac{1}{8}$ inch or 15 percent. (Other types of openings besides the flanged-in kind may also be installed in dished heads. The rules for such openings are given in PG-32 and discussed in chapter 6.) PG-29.3 also discusses multiple manholes in a head. Photo 3.2 shows a mud drum with a manhole ring.

Hemispherical Heads

The second type of dished head is the full hemispherical head. As the name indicates, the head is a complete hemisphere. The rules for calculating its thickness are given in PG-29.11. There are special considerations for hemispherical heads also:

1. A full hemispherical head may be thinner than the cylindrical shell to which it is attached. No other head may be thinner than the shell (PG-29.6).

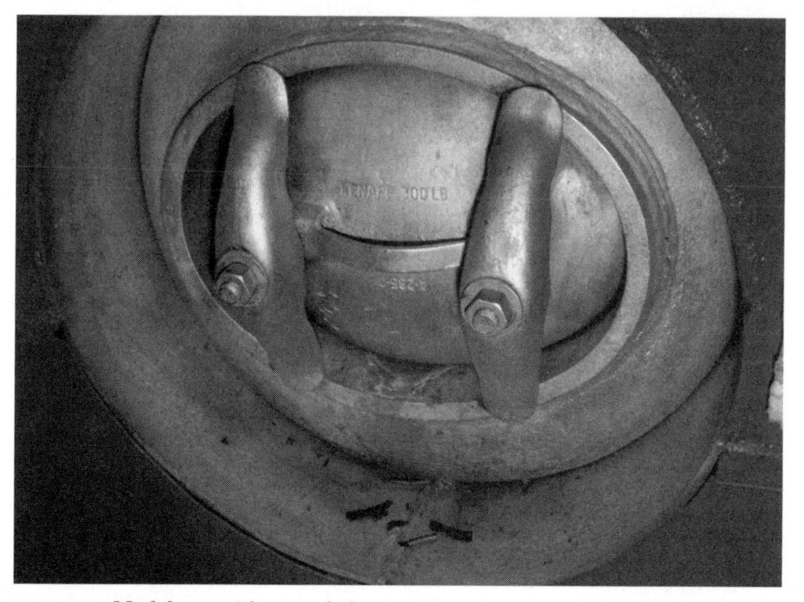

Photo 3.2 Mud drum with a manhole ring. Note the cover, made of SA 285, grade C, rated for 300 psi and supplied as a standard pressure part per PG-11.1.1. Photo by the authors.

2. No flange is required for attaching the hemispherical head to the shell (PW-13). Flanges are required for standard dished heads and ellipsoidal heads.

3. PG-29.11 gives formulae for both thin and thick heads. A hemispherical head is thick if the thickness calculated by the thin formula exceeds 35.6 percent of the inside radius. There are two thin formulae:

$$t = \frac{PL}{1.6S} \quad (1)$$

$$t = \frac{PL}{(2S)(0.2P)} \quad (2)$$

Considering the normal range of allowable stress values for boiler plate, one can see that for pressures less than roughly 20,000 psig, the second formula will call for a lower thickness. PG-29.11 instructs: "Use formula 1; however, formula 2 may be used for heads exceeding ½ inch in thickness that are to be used with shells or headers designed under the provisions of PG-27.2.2 and that are integrally formed on seamless drums or are attached by fusion welding. . . ." Since all cylindrical shells and headers are to be designed under the provisions of PG-27.2.2, and since the heads will be either integrally formed or attached by fusion welding, and since there is no longer any distinction in the Code between seamless and welded cylinders, it is unclear when, if ever, formula 1 is to be used. Guidance in this area may be found by examining older editions of Section I and Section VIII.

The 1949 edition of Section I gives the hemispherical head formula $t = PL/2SE$, with E being the joint efficiency. This is the same formula derived in chapter 2 for a thin spherical shell. By 1952, however, there were two thin hemispherical head formulae, then called A and B:

$$t = \frac{PL}{1.6SE} \quad (A)$$

$$t = \frac{PL}{2SE + 0.1} \quad (B)$$

Both formulae result in an increase in the calculated thickness for hemispherical heads. Formula A was intended for general use, and formula B was permitted under restricted circumstances, essentially the same as those listed previously for formula 2.

Section VIII, Division I, gives only one (thin) hemispherical head formula:

$$t = \frac{PL}{(2SE)(0.2P)}$$

This is the same as formula 2 from Section I, PG-29.11, when the efficiency $E = 1$.

Consideration of the history of the formulae for hemispherical heads in Section I; the wording of PG-29.11; and the formula in Section VIII, Division I, UG 32 (3) leads to the following conclusions regarding the use of formulae 1 and 2 in the current edition of Section I:

(a) Formula 1 (then formula A) was inserted into the Code during the time when riveted joints were common. This formula added conservatism to the calculation of required thickness for hemispherical heads to be attached by riveting.

(b) Formula 2 applies to all (thin) hemispherical heads that are more than ½ inch thick and either integrally formed on seamless cylinders or welded onto the cylinders. All such cylinders are now designed in accordance with PG-27.2.2, and there is really no case in which formula 2 may not be used, as long as the thickness is greater than ½ inch. It is recommended that the Code be changed to reflect the realities of modern construction and to bring it into line with Section VIII, Division 1 by eliminating formula 1 from future editions. Meanwhile, formula 2 should be used unless the thickness of the head computes to ½ inch or less. Only then should formula 1 be used.

4. PG-29.12 states that if a flanged-in manhole is to be installed in a hemispherical head, the thickness of the head will be the same as if the head were a standard dished head with an inside dish radius equal to 0.8 times the diameter of the shell and the greater of ⅛ inch or 15 percent added to the computed value of thickness.

5. PG-29.13 states that the thickness after forming of hemispherical (and ellipsoidal) heads shall be not less than that calculated by the appropriate formula.

Semiellipsoidal Heads

The third type of dished head in Section I is the semiellipsoidal head, defined in PG-29.7. The shape is a true ellipse with the depth of the head (which is half the minor axis) being equal to one-quarter the inside diameter of the head. This shape must be held to a tolerance of 0.0125 times the inner diameter (ID) of the head (PG-29.8). Semiellipsoidal heads are also called ellipsoidal heads. The rules for calculating the thickness are given in PG-29.7. A blank ellipsoidal head is calculated to be as thick as a cylindrical shell per PG-27.2.2. If a flanged-in manhole is to be installed, then the formula in PG-29.1 is used to calculate the thickness, using $L = 0.8$ times the shell diameter, and adding the greater of ⅛ inch or 15 percent to the calculated value. Semiellipsoidal heads require flanges per PW-13, 1 inch for diameters up to 24 inches and 1.5 inches for diameters over 24 inches.

These rules govern dished heads concave to pressure, sometimes called plus heads. Heads convex to pressure are designed by the rules in PG-29.9. The MAWP is equal to 60 percent of the value for a head of the same dimensions calculated by the formula in PG-29.1. The formulae in PG-29.7 and PG-29.11 (for hemispherical and ellipsoidal heads) specifically do not apply to heads with pressure on the convex side (minus heads).

Paragraph PG-30 gives rules for stayed dished heads. Stayed surfaces are discussed in chapter 5. It should be noted, however, that PG-30.2 gives a rule for calculating the largest permissible flat spot on a dished head concave to pressure. The diameter of the flat spot shall not exceed the allowable diameter for a flat head of the same thickness as given by the formula PG-31.3.2. (If the flat spot exceeds this diameter, then staying will be required.)

Sample Problems

Sample problem 3.1. Thickness of formed heads.

Find the required thickness for each type of blank head to fit a 60-inch-diameter drum if the MAWP is 400 psig. Assume the allowable stress is 15 ksi and the temperature does not exceed 650°F. For the standard dished head, assume that the radius of the dish is equal to the diameter of the shell.

Solution: For the standard dished head, use PG-29.1:

$$t = \frac{5PL}{4.8S}$$

where t = thickness in inches; P = MAWP, given as 400 psig; L = spherical or dish radius, given as 60 inches; and S = allowable stress at temperature, given as 15,000 psig.

Substituting, we get:

$$t = \frac{5(400)(60)}{(4.8)(15{,}000)}$$

$$= 1.667''$$

For the hemispherical head, use PG-29.11, formula 2:

$$t = \frac{PL}{2S - 0.2P}$$

where t = thickness in inches; P = MAWP, given as 400 psig; L = spherical radius, measured on the inside; and S = allowable stress, given as 15 ksi.

$$t = \frac{(400)(29.5)}{2(15{,}000) - 0.2(400)}$$

$$= 0.394''$$

Because the calculated thickness is less than ½ inch, we must use formula 1 from PG-29.11:

$$t = \frac{PL}{1.6S}$$

$$= \frac{(400)(29.5)}{1.6(15{,}000)}$$

$$= 0.492''$$

(Note that in using both formulae the inside radius is the outside radius minus the unknown thickness. The thickness was estimated at ½ inch to obtain $L = 30 - .5 = 29.5$ inches).

For the ellipsoidal head, use PG-29.7, which references PG-27.2.2:

$$t = \frac{PD}{2SE + 2yP}$$

where t = thickness in inches; D = the outside diameter of the flange of the head; S = allowable stress = 15 ksi; E = efficiency = 1.00; and y is a temperature-based coefficient = 0.4 (PG-27.4 Note 6).

$$t = \frac{(400)(60)}{2(15{,}000) + 2(0.4)(400)}$$

$$= 0.792''$$

Sample problem 3.2. Thickness of heads with a flanged-in manhole.

Solution: Given the same data as in sample problem 3.1, find the required thickness for each type of head if a 12 × 16-inch flanged-in manhole is to be installed in the head.

For a standard dished head, PG-29.3 indicates that an increase of the larger of ⅛ inch or 15 percent of the thickness calculated in problem 3.1 is required. Therefore the thickness is the larger of:

$$1.667'' + 0.125'' = 1.762''$$

or

$$1.667'' + 15\% \text{ or } 0.250'' = 1.917''$$

The answer, then, is 1.917 inches.

For a hemispherical head, PG-29.12 indicates that the thickness with a manhole is given by PG-29.1, using $L = 0.8D$ as indicated in PG-29.5:

$$t = \frac{5PL}{4.8S}$$

$$= \frac{5(400)(0.8)(60)}{(4.8)(15{,}000)}$$

$$= 1.333''$$

Then per PG-29.12 and PG-29.3, the thickness must be the greater of 1.333 inches + ⅛ inch or 1.333 inches + 15 percent, in which case t = 1.533 inches.

For an ellipsoidal head, per PG-29.7, 29.1, 29.5, and 29.3, the answer is the same for the ellipsoidal head as for the hemispherical head; so t = 1.533 inches.

Sample problem 3.3. MAWP on a head with a flanged-in manhole.

A dished head with a flanged-in manhole has a diameter of 48 inches and a dish radius of 38.4 inches. If the thickness is ⅞ inch, find the MAWP. The material is SA 285, grade C.

Solution: Use PG-29.1:

$$t = \frac{5PL}{4.8S}$$

where P = MAWP; L = dish radius of at least $0.8D$ = 38.4 inches; S = allowable stress per Section II, Part D = 13,800 psi; and t = thickness. But per PG-29.3, either ⅛ inch or 15 percent has been added to the calculated thickness because of the flanged-in manhole. So the actual thickness to be used in computing the MAWP will be the smaller of the two, found as follows:

$$\tfrac{7}{8}'' = t + \tfrac{1}{8}''; \; t = \tfrac{6}{8} = 0.75''$$

$$\tfrac{7}{8}'' = t + 0.15t; \; t = \frac{\tfrac{7}{8}''}{1.15} = 0.761''$$

The answer is 0.75 inch. Then:

$$0.75 = \frac{5P(38.4)}{(4.8)(13{,}800)}$$

$$= 259 \text{ psig}$$

We note that at $t = 0.833$ inch, 15 percent of the thickness equals ⅛ inch or 0.125 inch. Thus, in calculating the required thickness of a head with a flanged-in opening, if PG-29.1 gives an answer for thickness less than or equal to 0.833 inch, the amount to be added for the opening (per PG-29.3) will be ⅛ inch. If the thickness calculated by PG-29.1 is greater than 0.833 inch, then the amount to be added per PG-29.3 will be 15 percent of the calculated thickness from PG-1. Similarly, in calculating the MAWP on a head with a manhole or other flanged-in opening over 6 inches, either ⅛ inch should be subtracted from the actual thickness or the actual thickness should be divided by 1.15 to arrive at the t value used in PG-29.1 to compute the MAWP. If the actual thickness is less than or equal to 0.958 inch, then the t value will be the actual thickness minus ⅛ inch. If the actual thickness is greater than 0.958 inch, then the t value for formula PG-29.1 should be the actual thickness divided by 1.15.

Flat heads are calculated by relatively simple formulae, but there are terms in the formulae that depend on the way the heads are fastened to the vessel. Different ways of attaching the flat plate to the vessel provide different amounts of edge support to the plate. A number of the many methods of attaching flat heads and covers to vessels are illustrated in Figure PG-31 of the Code. Integrally formed head proportions are shown, as are the required weld sizes for heads and covers attached by welding.

Sample problem 3.4. Circular flat head per Figure PG-31(b).

Find the required minimum thickness for a flat circular cover as shown in Figure PG-31(b) if the diameter is 18 inches, the maximum allowable working pressure is 350 psig, and the allowable stress is 15 ksi.

Solution:

$$t = d\left(\frac{CP}{S}\right)^{1/2}$$

where t = thickness; d = diameter; P = MAWP; C is the factor from Figure PG-31(b) = 0.17; and S is allowable stress = 15,000 psi.

$$t = 18\left[\frac{(0.17)(350)}{15{,}000}\right]^{1/2}$$

$$= 1.134''$$

Sample problem 3.5. Noncircular flat head per Figure PG-31(g).

Find the required minimum thickness of a flat head of obround shape if the minor dimension is 12 inches and the major dimension is 18 inches. The head is inserted into the shell and attached by a full penetration groove weld, as shown in Figure PG-31(g). The maximum allowable working pressure is 400 psig and the allowable stress at temperature is 13.8 ksi.

Solution: PG-31.3.2(3) gives:

$$t = d\left(\frac{ZCP}{S}\right)^{1/2}$$

where t = required minimum thickness; d = the short dimension of the head = 12 inches; $Z = 3.4 - 2.4d/D = 3.4 - 2.4(12)/18 = 1.80$; P = MAWP = 400 psig; D = the long dimension of the head = 18 inches; S = allowable stress = 13,800 psi; and C = factor from Figure PG-31(g) = 0.33.

$$t = 12\left[\frac{(1.80)(0.33)(400)}{13{,}800}\right]^{1/2}$$

$$= 1.575''$$

Sample problem 3.6. Bolted circular cover with full face gasket per Figure PG-31(p) of the Code.

A 12-inch schedule 80 nozzle is to be fitted with a blind flange (i.e., a flat cover) attached with bolts and using a full-face gasket per PG-31(p). The bolt circle diameter is 21 inches. The maximum allowable working pressure is 850 psig, and the allowable stress for the cover plate material is 15 ksi. Find the required minimum thickness of the cover.

Solution: Use the equation:

$$t = d\left(\frac{CP}{S}\right)^{1/2}$$

where t = minimum thickness; d = bolt circle diameter per Figure PG-31(p); $C = 0.25$ per Figure PG-31(p); P = MAWP = 850 psig; and S = allowable stress at temperature (from Section II, Part D) for the cover material.

$$t = 21\left[\frac{(0.25)(850)}{15{,}000}\right]^{1/2}$$

$$= 2.499''$$

It must be noted that although the solution shown does meet the requirements of Section I, there are a number of design issues not covered. The choice of a bolt circle and the size, strength, and number of bolts are all important variables affecting the final design. It must also be emphasized that the full-face gasket design does not produce an edge moment and is therefore simpler to design. It may, however, require large bolt loads to compress the gasket sufficiently to produce a tight joint. It is therefore common to employ raised face flanges with relatively narrow gaskets and no metal-to-metal contact outside the gasket circle. See sample problem 3.7.

Sample problem 3.7. Bolted circular cover with a gasket and raised face blank flange with no metal-to-metal contact beyond the bolt circle (similar to Figure PG-31(j) of the Code).

A 12-inch NPS schedule 80 nozzle is to be fitted with a blind flange and bolted as shown in Figure PG-31(j). The bolt circle diameter is 19¼ inches, and there are twenty 1¼-inch bolts. The maximum allowable stress on the bolts is 15,000 psi. The gasket width w is ¼ inch, and the gasket circle mean diameter is 14 inches. The gasket compression factor m is 2.25, and the gasket seating stress y is 2200 psi. The blind flange is made of SA 105 material. Find the required minimum thickness of the cover if the MAWP is 850 psig.

Solution: Use PG-31.3.2(2):

$$t = d\left(\frac{CP}{S} + \frac{1.9Whg}{Sd3}\right)^{1/2}$$

where t = minimum required thickness; d = gasket diameter = 14 inches; C = 0.30; P = MAWP = 850 psig; S = allowable stress in cover material from Section II, Part D = 17,500 psig at 650°F; W = total bolt load; and hg = the distance from the centerline of the bolts to the centerline of the gasket = (19.25 − 14)/2 = 2.63. The total bolt load for operating conditions is the sum of the pressure loading on the cover and the force needed to keep compression on the gasket.

$$W = \frac{\pi}{4}d^2P + \pi d W_m P$$

$$= 0.785(14)2(850) + 3.14(14)(.25)(2.25)(850)$$

$$= 151{,}850 \text{ lb}$$

The total bolt load for gasket seating is taken as the average of the amount needed to seat the gasket and the amount available in the bolting. The gasket seating load is:

$$W = \pi(d)(w)(y) = 3.14(14)(.25)(2200) = 24{,}190 \text{ lb}$$

The available bolt load is the number of bolts times the root area of one bolt times the allowable stress on the bolting material. The root diameter for a 1¼-inch bolt having 7 threads per inch is 1.0725 inches. The root area is 0.903 square inch.

$$W = (20)(0.903)(15{,}000) = 271{,}023 \text{ lb}$$

The average of the bolt load for gasket seating and the available bolt load is (24,190 + 271,023)/2 = 147,606 lb.

The operating bolt load is the larger of the two values computed, so $W = 151{,}850$ lb. Then:

$$t = 14\left[\frac{0.3(850)}{17{,}500} + \frac{1.9(151{,}850)(2.625)}{17{,}500(14)^3}\right]^{1/2}$$

$$= 2.439''$$

Chapter

4

Cylindrical Parts Subject to External Pressure

Firetubes

The rules for designing firetubes, flues, and furnaces for firetube boilers are found in Part PFT of Section I and in Subpart 3 of Section II, Part D.

Firetubes, like watertubes, must be made of a material listed in PG-9. Further rules for firetubes are given in PFT-12, which directs the reader to PFT-50 and PFT-51.

PFT-50 gives the design temperatures to be used. PFT-51 gives the rules for design and specifies the use of external pressure charts found in Subpart 3 of Section II, Part D. The basis for the external pressure charts is given in Appendix 3 of Section II, Part D.

PFT-12.2 states that tubes may be attached by expanding and beading; expanding, beading, and seal welding; expanding and welding; or welding. Examples of the various attachment details are given in Figure PFT-12.1 of the Code.

PG-79 applies to tube holes for firetubes as well as watertubes. The holes may be drilled full-size, or they may be punched at least ½ inch undersize and then machined to the finish dimension. If the holes are thermally cut, then they must be cut sufficiently undersize that subsequent machining to the finish dimension will remove all the material with process-altered mechanical and/or metallurgical properties. PG-79 and PFT-12.2.5 both require that tube holes have their sharp edges removed before the tubes are installed.

Sample problem 4.1. Maximum allowable working pressure of a firetube (see Figures G, CS-1, and CS-2).

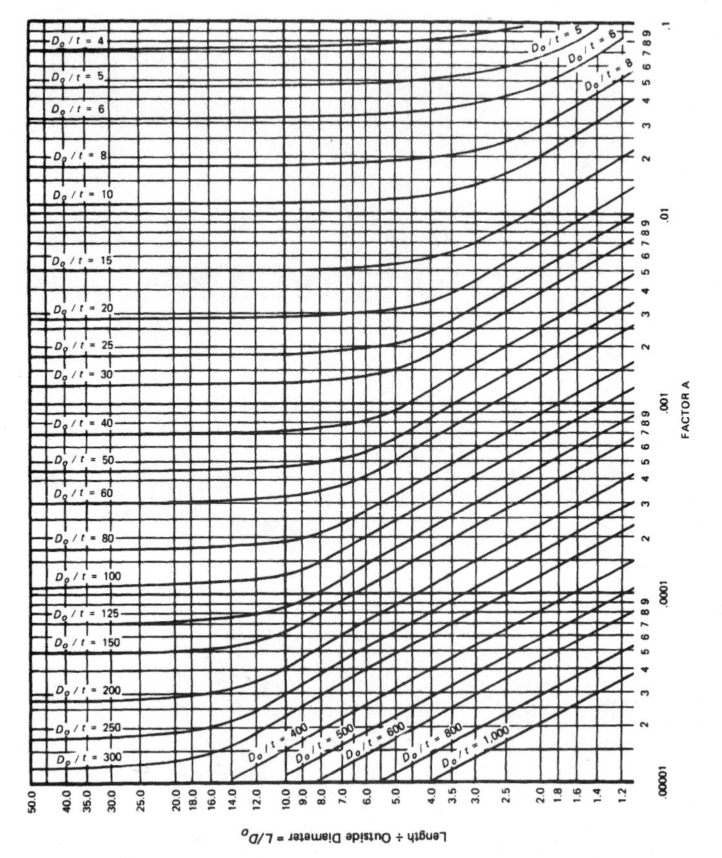

Figure G from Section II. Reprinted courtesy of ASME.

Find the MAWP on a firetube if the diameter is 3 inches and the thickness is 0.120 (11 gage). The material is SA-192.

Solution: PFT-12.1.1 states that the MAWP for a firetube is determined by the rules in PFT-50 and PFT-51. PFT-50.1 indicates that the design temperature for firetubes shall be as specified in PG-27.4, Note 2, the mean metal temperature, but not less than 700°F for a tube that absorbs heat. Section II, Part D indicates that for SA-192 (seamless tube), the appropriate external pressure chart is CS-1. The allowable stress at 700°F is 11.5 ksi.

The ratios L/D_o and D_o/t are calculated, where L is the tube length, D_o is the tube OD, and t is the tube thickness:

$$\frac{D_o}{t} = \frac{3}{0.120} = 25$$

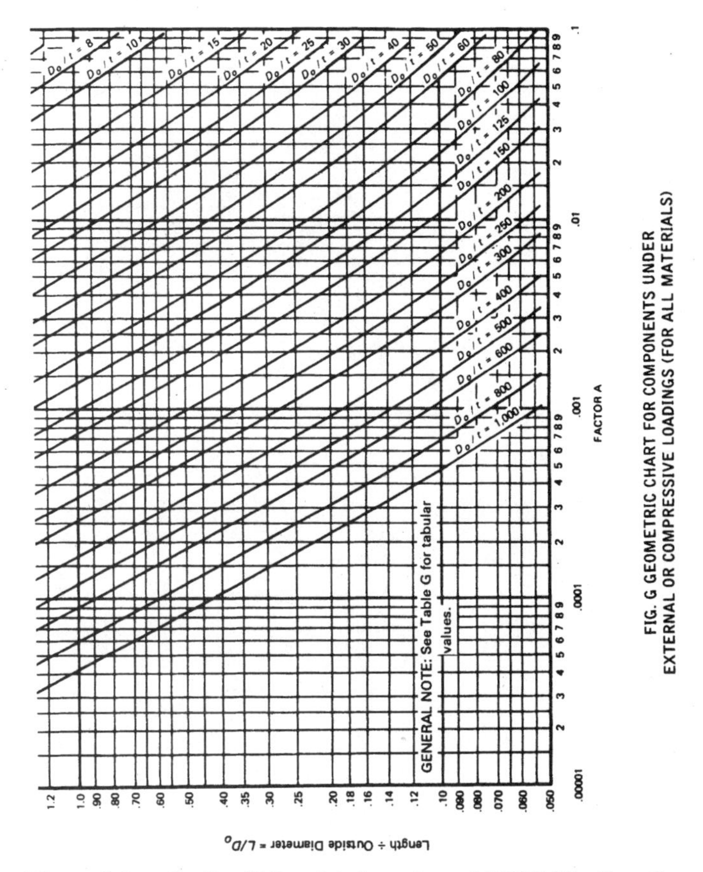

Figure G from Section II. Reprinted courtesy of ASME (*Continued*).

The tube length is not given, so it is assumed that the ratio L/D_o is greater than 50, corresponding to a tube length of at least 12½ feet. Using $D_o/t = 25$ and $L/D_o = 50$, Factor A is obtained from the geometric chart, Figure G, in Subpart 3 of Section II, Part D. (This procedure is used for D_o/t of 10 or greater.)

According to the chart, $A = 0.0018$. This value may be checked in the table, where it is found that for $D_o/t = 25$ and $L/D_o = 50$, Factor $A = 0.00176$.

Factor A is used to enter Chart CS-1, raising a vertical line to intersect the 700°F line. Factor B is then obtained by drawing a horizontal line from the intersection and reading the axis on the right. Factor $B = 7500$. This value may also be checked in the table, where it is found that for $A = 0.00176$ and $T = 700$°F, Factor B is midway between 7300 and 7700 psi. Interpolation is permitted, so $B = 7500$. The maximum allowable external pressure (Pa) on the tube is then calculated:

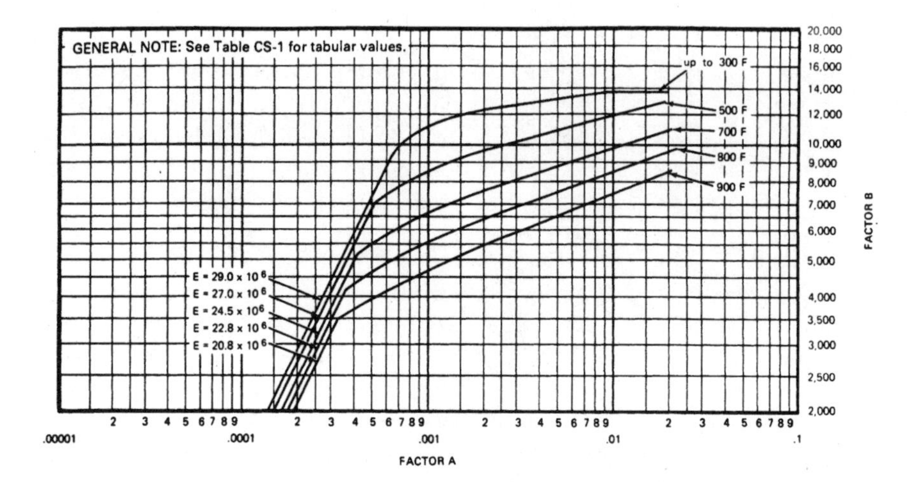

FIG. CS-1 CHART FOR DETERMINING SHELL THICKNESS OF COMPONENTS UNDER EXTERNAL PRESSURE WHEN CONSTRUCTED OF CARBON OR LOW ALLOY STEELS (SPECIFIED MINIMUM YIELD STRENGTH 24,000 psi TO, BUT NOT INCLUDING, 30,000 psi) [NOTE (1)]

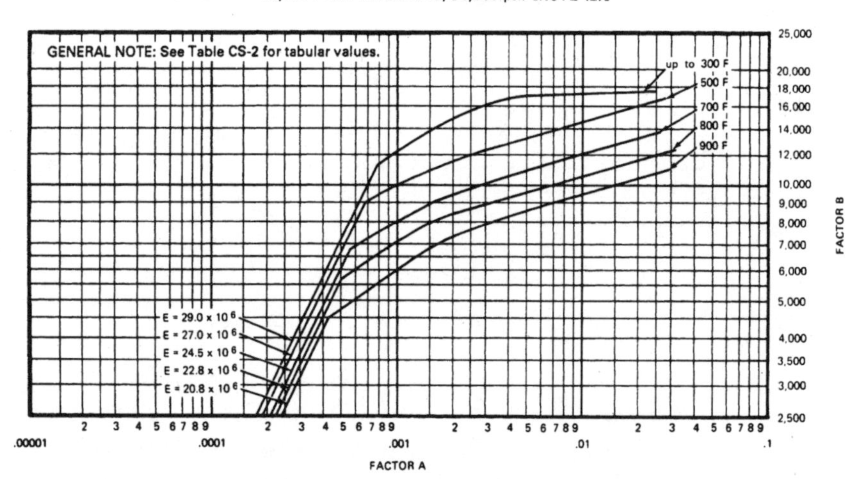

FIG. CS-2 CHART FOR DETERMINING SHELL THICKNESS OF COMPONENTS UNDER EXTERNAL PRESSURE WHEN CONSTRUCTED OF CARBON OR LOW ALLOY STEELS (SPECIFIED MINIMUM YIELD STRENGTH 30,000 psi AND OVER EXCEPT FOR MATERIALS WITHIN THIS RANGE WHERE OTHER SPECIFIC CHARTS ARE REFERENCED) AND TYPE 405 AND TYPE 410 STAINLESS STEELS [NOTE (1)]

Figures CS-1 and CS-2 from Section II. Reprinted courtesy of ASME.

$$Pa = \frac{4B}{3(D_o/t)}$$

$$= \frac{4(7500)}{(3)(25)}$$

$$= 400 \text{ psi}$$

We note that if the thickness is to be determined, the method is the same as in sample problem 4.1. A thickness is assumed and the corresponding maximum external pressure is calculated. If the calculated pressure is less than the required pressure (or if the calculated pressure greatly exceeds the required design pressure), a new thickness is assumed and the process is iterated until a satisfactory result is obtained.

Sample problem 4.2. Thickness of a firetube.

A small firetube boiler has tubes 10 inches long and 1⅜ inches in outside diameter. The tubes are SA-178A. The maximum allowable working pressure for the boiler is to be 150 psig. Find the minimum required thickness for the tubes.

Solution: $L/D_o = 10/1.375 = 7.27$. D_o/t must be guessed at (by estimating a thickness). Assume a 16-gage tube having $t = 0.065$ inch. $D_o/t = 1.375/0.065 = 21.15$.

Factor A is obtained from the geometric chart; $A = 0.0029$.

SA-178A is checked in Section II, Part D, and found to be on Chart CS-1. The temperature will be 700°F per PFT-50 and PG-27.4, Note 2. From the chart, $B = 8000$.

Then:

$$Pa = \frac{4B}{3(D_o/t)}$$

$$= \frac{4(8000)}{3(21.15)}$$

$$= 504 \text{ psi}$$

So per PFT-50.1.3 and PG-27.4, Note 8, $Pa = 510$ psi. This would be more than sufficient for the desired MAWP of the boiler. Corrosion and mechanical strength considerations would probably prevent going much thinner.

Furnaces

Many firetube boilers, including Scotch boilers and vertical firetube boilers, have furnaces made of boiler plate subjected to external pressure. The furnaces are of circular cross section, with firing taking place inside the furnace, which is surrounded and cooled by the water. The plate materials of which furnaces may be made are listed in PG-6.

The furnaces are subjected to external pressure. PG-80.2 states that cylindrical shells with a diameter of 24 inches or less that are subjected to external pressure may deviate from a true circle by not more than 1 percent of their outside diameter. For furnaces greater than 24 inches OD, the maximum permissible deviation from a true circle is calculated, based on the L/D_o and D_o/t ratios and the thickness of the furnace, using the chart in Figure PG-80.

PFT-14.1 states that the minimum thickness for a plain circular furnace is $\frac{5}{16}$ inch. Such furnaces may be seamless or welded. Welded furnaces require postweld heat treatment and either a bend test sample per PW-53 or a full X-ray of the long seam.

PFT-14.2 states that the MAWP on a plain circular furnace is computed per PFT-51.

Sample problem 4.3. Maximum allowable working pressure and maximum deviation from a true circle for a plain circular furnace.

A plain circular furnace is 22 inches OD and 6 feet long. The material is SA-285, grade C, and the thickness is $\frac{5}{16}$ inch. Find the MAWP on the furnace and find the maximum deviation from a true circle.

Solution: The maximum deviation from a true circle for a furnace not over 24 inches OD is given in PG-80.2(b) as 1 percent of the OD. Therefore the answer is $0.01(22) = 0.22$ inch.

The ratios are $L/D_o = 72/22 = 3.27$; and $D_o/t = 22/(\frac{5}{16}) = 70.4$.

Table 1A of Section II, Part D, indicates that for SA-285, grade C, the appropriate chart is CS-2. From Figure G, using $L/D_o = 3.27$ and $D_o/t = 70.4$, Factor A is found to be 0.00065.

To use chart CS-2, it is necessary to have a design temperature as well as a value for Factor A. PFT-50.1 indicates that for tubes, the design temperature shall be selected in accordance with PG-27.4, Note 2 (greater of mean metal temperature or 700°F), and for furnaces the design temperature shall be in accordance with PFT-17.7. PFT-17.7 gives the design temperature for the furnace as 100 degrees Farenheit above the water temperature (i.e., 100 degrees higher than the saturation temperature at the MAWP). Since the MAWP is not known and the water temperature is not given, an assumption must be made. Assume $T = 500$°F. On Chart CS-2 we note that Factor $A = 0.00065$ lies to the left of the temperature lines. We also note that the elastic modulus E is given as $27(10)^6$ psi at 500°F.

Then PFT-51.1.2 step 7 gives:

$$Pa = \frac{2AE}{3(D_o/t)}$$

$$= \frac{2(0.00065)(27)(10)^6}{3(70.4)}$$

$$= 166 \text{ psi}$$

We then check the steam tables for the saturation temperature at 181 psia. The saturation temperature is about 373.5°F. Therefore, the minimum design temperature per PFT-17.7 would have been 473.5°F, which means that the solution using 500°F is acceptable.

We note that the Code is ambiguous as to whether rounding up to 170 psi would be allowed on this furnace. PFT-50.1.3 states: "Rounding off equation results to the next higher unit of 10 is permitted. (See PG-27.4, Note 8.)" Note 8 refers to a "pipe of a definite minimum wall thickness" and in other contexts is understood to apply to tubes and pipes but not shells. It is not clear whether the intent is to permit rounding up generally or only for tubes.

Sample problem 4.4. Thickness and out-of-roundness of a furnace.

A furnace is made of SA-515 grade 60 material for a maximum external pressure of 200 psig. The furnace is 36 inches in diameter and 8 feet long. Find the required minimum thickness. Find also the maximum permissible out-of-roundness.

Solution: A thickness is assumed: $t = 3/8$ inch. The ratios L/D_o and D_o/t are computed: $L/D_o = 8(12)/36 = 2.67$; and $D_o/t = 36/(3/8) = 96$.

Table 1A of Section II, Part D gives external pressure chart CS-2 for SA-515 grade 60. Then from Figure G in Subpart 3 of Section II, Part D, Factor $A = 0.00036$. The steam tables show a saturation temperature of 388°F at 215 psia. Our design temperature must be at least 488°F. For covenience we use the 500°F line on Chart CS-2. Factor $B = 4700$.

Now apply PFT-51.1.2 step 6:

$$Pa = \frac{4B}{3(D_o/t)}$$

$$= \frac{4(4700)}{(3)(96)}$$

$$= 65 \text{ psig}$$

Not thick enough. Try $t = 1/2$ inch: $D_o/t = 72$. $A = 0.0008$. $B = 9400$.

$$Pa = \frac{4(9400)}{3(72)}$$

$$= 174 \text{ psig}$$

Still too thin. Try $9/16$ inch thickness. $D_o/t = 64$. $A = 0.00092$. $B = 9900$.

$$Pa = \frac{4(9900)}{3(64)}$$

$$= 206 \text{ psig}$$

The maximum out-of-roundness is found for diameters greater than 24 inches in PG-80.1 and Figure PG-80.1. For $L/D_o = 2.67$ and $D_o/t = 64$, Figure PG-80.1 gives $e = 0.45t$:

$$e = 0.45(9/16)$$

$$= 0.25''$$

Corrugated furnaces

As indicated earlier, shells subject to external pressure are subject to failure by collapse and are accordingly rated for a lower pressure for a given thickness than shells subject to internal pressure. For example, the furnace of sample problem 4.4 was rated for an MAWP of 206 psig external pressure. Calculated by the formula in PG-27.2.2 (for internal pressure), the shell would have an MAWP of 475 psig. The resistance of cylindrical shells to failure by buckling collapse under external pressure can be increased by using corrugated rather than plain furnaces. Such corrugated furnaces must be made from PG-6 materials. The requirements for five such furnaces are given in PG-18. These requirements include the specifications for the corrugations for each of the five types. The plain circular ends must not exceed 9 inches in length.

PG-18.1.1 indicates that the longitudinal and circumferential welds may be of the double-welded butt type. Postweld heat treatment is required. The longitudinal seams may be accepted by a bend test on a sample per PW-53, in which case no radiography is required, or the long seam may be fully radiographed, in which case the bend test for each section is not required (PFT-18.1.2). The thickness of the corrugated furnaces must be measured, and PFT-18.2 contains instructions as to where the measurement shall be made for each type of furnace. If the measurement is made by drilling a hole, it must be a ⅜-inch hole, to be later closed with a plug and located on the bottom of the furnace. (One would expect such measurements to be made ultrasonically on any corrugated furnaces manufactured currently.)

The maximum allowable pressure on a corrugated furnace is given by PFT-18.1:

$$P = \frac{Ct}{D}$$

where P = MAWP; C = a constant depending on the type of furnace, given in PFT-18.1; t = furnace thickness, not less than 5/16 inch for any furnace and not less than 7/16 inch for the Purves type furnace; and D = mean diameter of the furnace.

Sample problem 4.5. MAWP of a corrugated furnace.

Find the MAWP on a Morison-type corrugated furnace if the thickness is $\frac{5}{16}$ inch and the least diameter is 26 inches.

Solution: $P = Ct/D$; $C = 15{,}600$; $t = \frac{5}{16}$ inch; and $D = 28$ inches per PFT-18.1. Therefore:

$$P = \frac{15{,}600(\frac{5}{16})}{28}$$
$$= 174 \text{ psig}$$

Combined Plain and Corrugated Furnaces

The Code also allows the use of furnaces made by combining a section of plain circular furnace with a section of corrugated furnace (PFT-19). Each of the furnaces must be independently self-supporting; neither can reinforce the other. The corrugated section of the furnace is designed per PFT-18.1 (PFT-19.3), and the plain circular section is designed per PFT-51, but for twice the actual section length (PFT-19.2). The weld joining the two sections is shown in Figure PFT-19. It must be a full penetration butt weld located not more than the lesser of 1½ inches or three times the corrugated furnace thickness from the point of tangency to the first corrugation. Both furnace sections require PWHT, and the long seams of each section must have either a bend test plate or full radiography.

Adamson ring furnaces (photo 4.1)

The resistance to external pressure of plain circular furnaces may be increased by effectively reducing the length of the furnaces between supports. This can be done by applying external stiffening rings to the furnace sections. One specific ring reinforcement design dating from the days of riveted boilers is the Adamson furnace, defined and described in PFT-16. This design does not require any calculations for the ring reinforcement; the furnace is designed per PFT-51 using the distance between Adamson rings as the length. No other calculations are required, provided the furnace and ring meet the detailed requirements of PFT-16 and Figure PFT-16. The section length must be at least 18 inches, and the minimum thickness is $\frac{5}{16}$ inch. The Adamson ring welds require PWHT, as do the longitudinal and circumferential seams of the furnace sections. The longitudinal seams of the sections also require either bend samples or a full X-ray.

Ring reinforced furnaces

Circular furnaces may also be reinforced by the addition of welded rings spaced along the external length. The stiffening rings must go

Photo 4.1 A large Adamson furnace at the Adamson Works. Photo courtesy Doug Hague, LR Insurance.

completely around the furnace (PFT-17) and must be rectangular in cross section. They must be full-penetration-welded to the furnace (PFT-17.1). The required proportions of the rings are given in PFT-17.2 to PFT-17.6. Other design requirements are given in PFT-17.7 to PFT-17.11, and a ring furnace sketch is given in Figure PFT-17.2. As with other furnaces, the longitudinal and circumferential seams require PWHT (as do the welded rings), and the long seams require either bend test samples or full X-ray. The following sample problem illustrates the design of a ring reinforced furnace.

Sample problem 4.6. Ring reinforced furnace.

A plain circular furnace is 32 inches in outside diameter and 12 feet long. The thickness is ½ inch and the material is SA-285, grade A. Assume the design temperature to be 500°F. Then: Find the MAWP on the furnace. Assume reinforcing rings are added to the furnace spaced on 3-foot centers. Find the MAWP for the furnace. Assume the rings will be made of SA-285 A also. Find an acceptable height and width for the rings.

Solution: (a) From Section II, Part D, SA-285 A is on Chart CS-1 and has allowable stress equal to 11,300 psi for T up to 650°F. $D_o = 32$ inches. $L = 12(12) = 144$ inches. $t = \frac{3}{8}$ inch. Therefore, $L/D_o = 144/32 = 4.5$; and $D_o/t = 32/(\frac{3}{8}) = 85.33$.

Figure G gives Factor $A = 0.000032$. Chart CS-1 is used to find Factor B, but a temperature is needed to enter the chart. Assume $T = 500$°F. $B = 4600$.

The MAWP is given by PFT-51.1.2, step 6:

$$Pa = \frac{4B}{3(D_o/t)}$$

$$= \frac{4(4600)}{3(85.33)}$$

$$= 72 \text{ psig}$$

Again we note that rounding up to 80 psig appears to be allowed by PFT-50.1.3. This is, however, a 10 percent increase and would not be recommended.

(b) The allowable pressure on the furnace with reinforcing rings every three feet is found the same as in (a), but with $L = 36$ inches. Thus, $L/D_o = 36/32 = 1.13$; and $D_o/t = 32/(3/8) = 85.33$. Factor $A = 0.0014$. Factor $B = 9050$.

$$Pa = \frac{4B}{3(D_o/t)}$$

$$= \frac{4(9050)}{3(85.33)}$$

$$= 141 \text{ psig}$$

The MAWP on the furnace can be nearly doubled by the addition of the reinforcing rings. The rings remain to be designed.

(c) The requirements for the reinforcing rings are given in PFT-17. PFT-17.1 requires that the rings be either continuous or formed into a full circle by welding together plate sections or bars using full penetration welds. We assume our rings will be so fabricated.

PFT-17.2 requires that the ring thickness be not less than five-sixteenth and not more than the lesser of thirteen-sixteenth or 1.25 times the furnace thickness.

PFT-17.3 requires that the height to thickness ratio for the rings be between 3 and 8 (inclusive).

PFT-17.4 requires that the ring be attached to the furnace by use of full penetration welds. We so specify for our rings in this case.

PFT-17.5 requires that the furnace thickness be at least 5/16 inch. Our furnace is 3/8 inch thick.

PFT-17.6 requires that the ring spacing not exceed 36 inches. We are using 36-inch spacing.

PFT-17.7 states that the design temperature for the furnace shall be 100°F higher than the water temperature. The steam tables show a saturation temperature of 362°F for an absolute pressure of 141 + 15 = 156 psi. The design temperature would therefore be at least 462°F, so our assumption of 500°F is satisfactory.

PFT-17.8 requires the design of the boiler to permit replacement of the furnace. This requirement recognizes that the thinner furnace permitted by the use of the reinforcing rings will be more vulnerable to corrosion. Therefore, the furnace must be capable of being disconnected at the tubesheets and slid out of the boiler. The rings must not obstruct the removal of the furnace.

PFT-9 requires postweld heat teatment of the completed furnace. This requirement includes the attachment welds to the tube sheets as well as the circumferential, longitudinal, and ring welds. There is also a statement that radiography is not required, but this applies only to the circumferential joints attaching the furnace to the tubesheets (see PFT-17.11.2 and 17.11.3, described below).

PFT-10 requires the use of PFT-51 to determine the MAWP (as we have already done).

PFT-11.2 and 11.3 require that each furnace section longitudinal seam have either a bend test sample or full radiography. (This requirement is the same for other furnace types.)

PFT-11.1 gives the method for calculating the required moment of inertia of the ring cross section. It should be noted that the formula for the actual moment of inertia of the ring is not given in the Code. $I = bH^3/12$, where b = the ring thickness and H = the ring height. The method is trial and error. A ring is selected and its moment of inertia is calculated. The required moment of inertia is calculated by the methods of PFT-51.11.1, and the required moment of inertia is compared to the actual moment of inertia. If necessary, another ring section is selected and the computations repeated until a satisfactory solution is obtained.

First a ring section is chosen. It is suggested that the largest possible ring section be chosen for the first try. The largest width is either $^{13}/_{16}$ inch or 1.25 times the furnace thickness. 1.25(.375) = $^{15}/_{32}$ inch = 0.4688 inch. This is less than $^{13}/_{16}$ (or 0.8125 inch), so the maximum thickness will be $^{15}/_{32}$ inch. The maximum height is eight times the thickness, so the height will be $8(^{15}/_{32})$ = 3.75 inches. The ring cross section properties are:

$$\text{Area } As = (^{15}/_{32})(3.75) = 1.758 \text{ in}^2$$

$$I = bH^3/12 = (^{15}/_{32})(3.75)^3/12 = 2.060 \text{ in}^4$$

The required cross section is then found: Use PFT-17.11.1, step 1, to calculate Factor B:

$$B = \frac{PD_o}{t + (As/Ls)}$$

where Ls is the length between sections, center to center, if the sections are symmetrically placed. (See PFT-51.1.1 for the definitions.)

$$B = \frac{141(32)}{^3/_8 + (1.758/36)}$$

$$= 10{,}646$$

The calculated value of B is used with a materials chart in Section II, Part D, Subpart 3, to obtain a value for Factor A. The chart to be used is the chart for the ring material, and the temperature to be used is the furnace temperature. In this case we use Chart CS-1 and 500°F. We enter the chart along the axis on the right at $B = 10{,}646$ and then move horizontally to the left to the 500°F temperature line. The value of A is then found by moving vertically down to the axis at the bottom of the chart. In this case, $A = 0.0042$.

The value of Factor A obtained from the chart is then used (PFT-51.1.1, step 5) to compute the required moment of inertia Is:

$$Is = \frac{D_o^2 Ls[t + (As/Ls)]A}{14}$$

$$= \frac{(32)^2(36)[0.375 + (1.758/36)](0.0042)}{14}$$

$$= 4.688 \text{ in}^4$$

This value for the required moment of inertia is substantially greater than the actual moment of inertia for the chosen ring. We have already chosen the largest ring section permitted by the design rules of PFT-17, so we conclude that we cannot meet the Code with a ring spacing of 36 inches. We must decrease the ring spacing and recompute. We assume the pressure remains at 141 psig and the ring dimensions remain the same. Let the spacing be 24 inches.

Then:

$$B = \frac{141(32)}{(3/8 + 1.758/24)}$$

$$= 10{,}066$$

Chart CS-1 gives $A = 0.0027$.

$$Is = \frac{(32)2(24)(3/8 + 1.758/24)(0.0027)}{14}$$

$$= 2.125 \text{ in}^4$$

Much better, but still not good enough. The calculation is repeated for $Ls = 22$ inches. The required moment of inertia is 1.98 inches. We therefore conclude that we can use the furnace for 141 psig if we use six reinforcing rings on 20.57-inch centers.

Sample external pressure problems are also given in the appendix of Section I, A-381, A-382, and A-383.

Chapter

5

Stays, Staybolts, and Stayed Surfaces

In chapters 2, 3, and 4, the rules for designing the many boiler components of cylindrical and/or spherical shape were discussed and illustrated. These components were self-supporting under the applied pressure loading because of their shapes and thicknesses. In chapter 4 we also discussed self-supporting flat heads and covers. Such covers must be relatively thick in order to carry the pressure without excessive deformation. Many boiler designs require the use of flat components that carry pressure on one side; these components can be made much thinner if the flat plates can be supported at points on the surface as well as at the edges. Similarly, there are conditions when even curved components can be appreciably thinner if intermediate support can be provided. (One example of this already discussed is the ring-reinforced cylindrical furnace.)

The predominant method of providing intermediate reinforcement to flat surfaces is to attach tension members to points on a grid on the surface subjected to pressure. The tension members are called stays. The grid pattern may be square (square pitch), rectangular, or radial and circumferential. In any case the center to center distance between the stays is the pitch. Stays called through stays run the full length of a boiler, from one head to the other. Diagonal stays support flat heads and run diagonally to points of attachment on the shell. Girder stays support the top plate in combustion chambers. Many boiler designs contain water legs, formed by flat plates separated from each other by a few inches. The plates are joined together by staybolts, and thus support each other.

The rules for stayed surfaces and stays are found in a number of different areas of the Code, as listed and summarized below:

The relevant passages are PG-13, 46, 47, 48, 49, 82, and 29; PW-12 and 13; PFT-22, 23, 24, 25, 26, 27, 28, 29, 30, 31, and 32; A-8 and 10; Table A-4; and Figure A-8.

PG-13 requires that stays be made of material that complies with SA-36 or SA-675.

PG-46 gives rules for the plate that is to be supported by stays.

PG-47 gives rules for staybolts and references PW-19 for welded staybolts.

PG-48 gives the rule for the maximum distance from the edge of a flanged flat head to the first row of stays. This rule is illustrated in the Appendix in Figure A-8, (i) and (j). The rule for an unflanged head is illustrated in Figure A-8 (k).

PG-49 gives dimensional requirements for staybolts.

PG-82 gives the requirements for holes for threaded stays. The holes for threaded stays are drilled full-size or punched and then drilled or reamed to size. For $\frac{5}{16}$-inch plate, the punched hole must be at least $\frac{1}{8}$ inch smaller than the finished hole diameter. For plates over $\frac{5}{16}$-inch thickness, the punched holes must be at least $\frac{1}{4}$ inch less than the final diameter. Threaded holes "shall be tapped fair and true, with a full thread."

PW-29 gives rules for the holes for welded stays.

PWT-12 gives rules for staybolts and welds on box headers and welded waterleg joints, and for the pitch of supporting staybolts.

PWT-13 gives rules for areas of heads to be stayed.

PFT-13.1 gives the rule for calculating the pressure (or thickness) of a tubesheet in a firetube boiler where the tubesheet carries the ends of the girder stays.

PFT-13.2 and 13.3 allow and regulate the use of sling stays in place of girder stays.

PFT-22 incorporates the rules from PG and PW, as applicable, to firetube boilers.

PFT-23 gives rules for curved stayed surfaces, allowing some credit to be taken for the inherent strength imparted by the curved shape. Special boiler types, including locomotive and vertical firetube, are discussed.

PFT-24 gives rules and formulae for computing the areas of (flat) heads to be stayed. PFT-24.4 gives special rules for through stays in HRT boilers.

PFT-25 and Figure PFT-25 give rules for determining when segments of heads require staying.

PFT-26 gives the general rule for computing the magnitude of the load carried by each stay. It is emphasized that this paragraph is the basis for the method used to compute all stays and staybolts, in whatever type of boiler.

PFT-27 gives many rules for the spacing of stays at the edges of areas to be stayed.

PFT-28 gives rules for calculating the allowable stress on stays and staybolts (referencing PG-49).

PFT-29 gives rules for flexible staybolts.

PFT-30 gives rules for girder stays.

PFT-31 gives rules for stay tubes.

PFT-32 gives rules for diagonal stays.

Formulae for computing root diameters of threaded stays and staybolts are given in A-8.

Areas of flanged heads to be stayed are given in A-10 and Table A-4. For heads of certain preferred proportions, the area to be stayed may be read directly from the table instead of being calculated per PFT-24.

Figure A-8 gives a number of sketches illustrating the application of the rules for locating stays.

The applications of the foregoing rules are discussed below and illustrated in sample problems.

PG-46.1 gives the formula for finding the required minimum thickness of a plate to be stayed; the MAWP on a plate; or the required maximum pitch of the stays on a plate of given thickness carrying a given MAWP. The plate must be strong enough to resist deformation between the stays. The formula is:

$$t = p\left(\frac{P}{SC}\right)^{1/2}$$

or

$$P = \frac{t^2SC}{p^2}$$

where t = minimum thickness required for the plate that is to be supported by the stays; p = *maximum* pitch of stays, if the pitch is not equal in all directions; P = MAWP for the stayed plate; S = allowable stress for the plate material (PG-46.1 is concerned with the strength of the plate, not of the stays); and C is a factor varying from a low value of 2.1 to a maximum value of 3.2. The value of C for a particular case is determined by the amount of support the plate receives from the type of attachment of the stays to the plate. Five different cases (i.e., 5 values of C) are given in PG-46.1.

PG-46.2 to PG-46.8 give additional rules for stayed surfaces (e.g., the minimum thickness of a plate to be stayed is $\frac{5}{16}$ inch). The maximum pitch for threaded staybolts is $8\frac{1}{2}$ inches, and for welded staybolts it is 15 times the diameter of the staybolt.

PG-47 contains rules for threaded staybolts (PG-47.1 and PG-47.2) and indicates that the requirements for welded staybolts are found in PW-19 (see sketch 5.1). Threaded staybolts are installed in screwed holes in the plates. The holes are threaded by use of a special long staybolt tap, which causes the threading to be a continuous helix in both plates. This permits the staybolt to be threaded first through one plate and then through the other. The staybolt must extend past the outer surface of each plate at least two full threads, and its ends are riveted over. Alternatively, the staybolts may have threaded nuts. In this latter case the staybolts must extend fully through the nuts. Staybolts that are 8 inches or less in length must have tell-tale holes drilled in their ends. These holes must be at least 3⁄16 inch in diameter and extend into the staybolt to a depth at least ½ inch beyond the inner surface of the plate to be stayed. (The staybolts may also be hollow.) The reason for the tell-tale holes is to detect cracking of the staybolts. Typically, one plate is relatively cold, while the opposite plate is exposed to the fire and is therefore much hotter. Differential thermal expansion causes bending stresses on the short staybolts. Additionally, the threads cause stress concentrations at the roots, particularly adjacent to the plates. Leakage at the tell-tale holes indicates that there is cracking of the staybolt. (The other traditional method for checking staybolts for cracking is to strike the ends with an inspector's hammer and listen for the ring of a sound staybolt or the dull thud of a cracked one.)

Welded staybolts, staybolts over 8 inches long, and ball and socket or jointed staybolts do not require tell-tales, except that, in the case of threaded staybolts in the waterlegs of watertube boilers, the staybolts must be either hollow or drilled at both ends, regardless of length.

PFT-26 states that the area used in computing the load on a stay shall be the full pitch area (i.e., either p^2 for a square pitch or p_1p_2 for rectangular pitch), with the area of the stay deducted. It is not clear from the Code whether the outside diameter or the root diameter of a

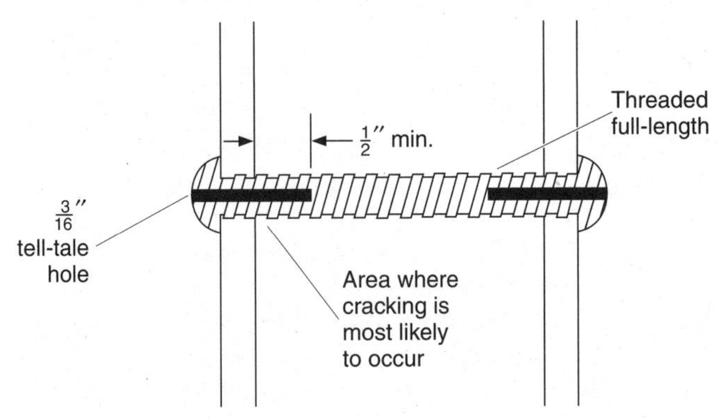

Sketch 5.1 Threaded staybolt with tell-tale holes (may also be hollow).

threaded stay should be used in computing the area of the stay for purposes of PFT-26. It is recommended that you deduct the root area of the threaded stays to obtain the net area as required by PFT-26. It is slightly conservative to follow this practice, and the root area has to be calculated anyway for the staybolt or stay stress calculations. The load to be carried by one stay is therefore:

$$\text{load} = P\left(p^2 - \frac{\pi d^2}{4}\right)$$

or

$$\text{load} = P\left(p_1 p_2 - \frac{\pi d^2}{4}\right)$$

PG-49 references PFT-26 as the basis for computing the load on a staybolt and indicates that the area of a staybolt should be taken at the least diameter. The required area is to be the load on the staybolt divided by the allowable stress on the staybolt material (per Section II, Part D, Table 1A), multiplied by 1.1. Computationally, this is the same as dividing the allowable stress from the table by 1.1, and it is generally most convenient to perform this division at the outset in staybolt and stay problems. PFT-28.1 states that the required area for staybolts *and stays* shall be computed per PG-49. The area of any tell-tale holes must be added to the required area computed by the rule of PG-49 (see sketch 5.2). The least diameter of tell-tale holes is $\frac{3}{16}$ inch, in order that they do

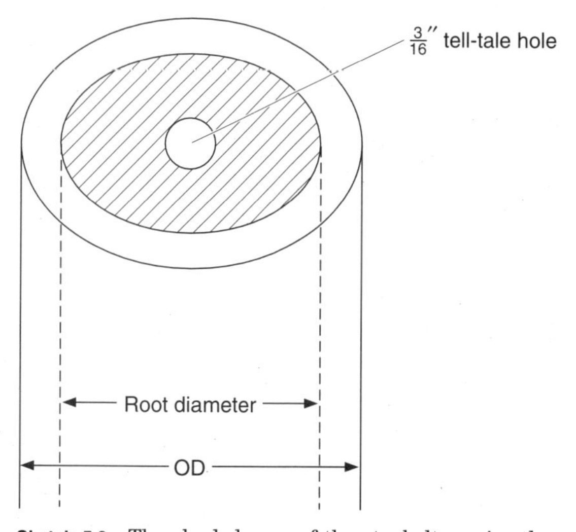

Sketch 5.2 The shaded area of the staybolt carries the load.

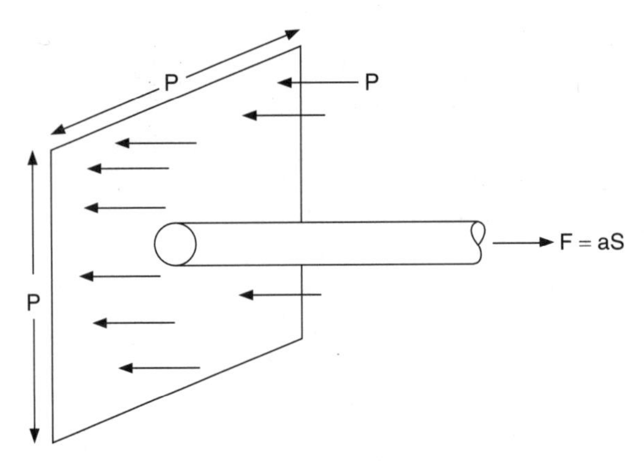

Sketch 5.3 Basic staybolt equation: $PA = naS$, where S = allowable stress from Table 1A divided by 1.1.

not become plugged so they fail to perform their function. There is no advantage to the manufacturer in making them larger, so tell-tale holes are generally 3⁄16 inch. The area of a 3⁄16-inch tell-tale hole is 0.0276 in^2.

Using the methods for computing the loads and areas discussed above we can state the basic staybolt equation (see sketch 5.3):

$$PA = na\left(\frac{S}{1.1}\right)$$

where P = maximum allowable working pressure; A = the net area subjected to pressure loading = the pitch area minus the area of stay(s); n = the number of staybolts or stays (must be an integer); a = the net area of the staybolt or stay = the root area minus 0.0276 for threaded staybolts; and S = allowable stress on the staybolt or stay, from Section II, Part D, Table 1A.

We note that all the pressure load on the plate is considered to be carried by the stay. There are no shear stresses at the edges of the pitch area, and no other support is available except at the edges of the plate. Each stay can be likened to a kite string with the area supported considered as the kite. The full pressure load is carried by the string, but the kite must be stiff enough to carry the load to the string. Photos 5.1 and 5.2 show stayed boilers.

Sample Problems

Sample problem 5.1. Maximum pressure on a stayed plate.

Find the maximum allowable working pressure on a plate having a thickness of 3⁄8 inch if it is supported by stays on a square pitch of 8½

Photo 5.1 A locomotive boiler under test. Note the riveted construction and the stayed water-legs of the firebox at the left end. Courtesy Doug Hague, LR Insurance.

inches. Assume the allowable stress in the plate is 13,800 psi. The stays are screwed staybolts with the ends riveted over.

Solution: PG-46.1 gives:

$$P = \frac{t^2SC}{p^2}$$

where P = MAWP; $t = \frac{3}{8}$ inch; S = 13,800; C = 2.1 (screwed stays in plate not over $\frac{7}{16}$ inch); and $p = 8\frac{1}{2}$ inches.

Substituting, we get:

$$P = \frac{(\frac{3}{8})^2(13{,}800)(2.1)}{(8.5)^2}$$

$$= 56 \text{ psig}$$

Sample problem 5.2. Maximum pitch of stays.

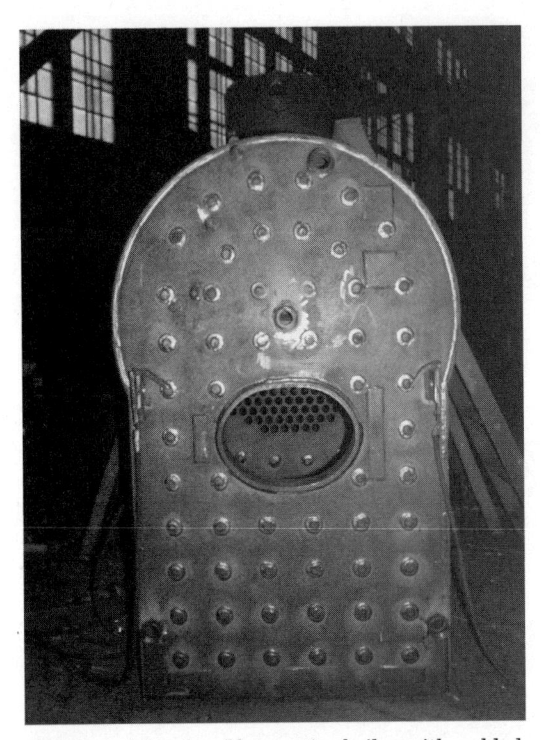

Photo 5.2 A "modern" locomotive boiler, with welded staybolts. Note the firing door ring and the tubesheet awaiting the installation of tubes. Photo by the authors.

Assume all data from sample problem 5.1, but a maximum allowable working pressure of 125 psig is desired. Find the maximum square pitch. Find also the maximum rectangular pitch.

Solution: From PG-46.1, formula 1:

$$p = \frac{t}{(P/SC)^{1/2}}$$

$$= \frac{0.375}{125/[(13{,}800)(2.1)]^{1/2}}$$

$$= 5.71''$$

We note that the maximum square pitch is 5.71 inches. Further, PG-46.1 defines the pitch p as the maximum pitch measured between straight lines through the centers of the stays. Therefore, if the pitch in this case is to be rectangular, its largest dimension will be 5.71 inches. Its least dimension could be any practicable value less than 5.71 inches.

Sample problem 5.3. Size of staybolts.

A plate is to be stayed on a 6-inch (square) pitch. The maximum allowable working pressure is 175 psig. The staybolts will be threaded and riveted over. The thread will be a 12 V thread. Find the smallest standard staybolt size for this plate. The plate is SA-515 grade 60. The staybolts are SA-36.

Solution: For the staybolt, use:

$$PA = naS$$

where $P = 175$; $A = p^2 - ar$ (PFT-26.1); $n = 1$; a = net area = ar – tell-tale hole area (PG-49.2 and PFT-28.1) = $ar - 0.0276$; S = allowable stress on the staybolt/1.1 (PG-49.1) = 12,600/1.1, where the allowable stress is obtained from Section II, Part D, Table 1A; ar = root area of the staybolt; d = root diameter of the staybolt; D = nominal diameter of staybolt; and P = thread pitch = 1/number of threads per inch.

$$175(36 - ar) = (1)(ar - 0.0276)(12{,}600/1.1)$$

$$ar = 0.569''$$

Then:

$$ar = \frac{\pi d^2}{4}$$

$$0.569 = \frac{\pi d^2}{4}$$

$$d = 0.851''$$

Therefore, from Appendix A-8:

$$d = D - 1.732P$$

$$0.851 = D - 1.732(\tfrac{1}{12})$$

$$D = 0.995''$$

Staybolts are generally available in increments of $\frac{1}{16}$ inch. In this case the nearest standard size would be 1 inch.

Sample problem 5.4. Area of a head to be stayed.

A horizontal return tube boiler is 60 inches in diameter and has a maximum allowable working pressure of 200 psig. The head thickness is $\frac{9}{16}$ inch, and the material is SA-285 grade C. The distance from the tops of the top row of tubes to the shell is 20 inches. The head has an outer radius of flange equal to 3 inches. Find also the maximum distance from the shell to the first row of braces.

Solution: The area to be stayed is illustrated in Figure PFT-24.1 and given in PFT-24. The area is given by formula PFT-24.3:

$$A = (4/3)(H - d - 2)^2 \left[\frac{2(R - d)}{(H - d - 2) - 0.608} \right]^{1/2}$$

where H = distance from tube to shell; A = area to be stayed; d = distance, from PFT-24.1; and R = radius of boiler.

PFT-24.1 stipulates that d *may* be the larger of the following: (a) the outer radius of the flange, but not more than eight times the head thickness; or (b) $80t/(P)^{1/2}$. Therefore:

$$d = 3'', \text{ which is less than } 8(9/16) \text{ or } 4.5''$$

or

$$d = \frac{80(9/16)}{(200)^{1/2}} = 3.182$$

We may use either $d = 3$ inches or $d = 3.182$ inches in determining the area to be supported. If $d = 3$ inches is chosen, the area may be found (A-10) in Table A-4 directly. Using this table, we see that for a 60-inch diameter boiler with 20 inches above the tubes, the area to be stayed may be read as 519 in^2. Alternatively, we may use $d = 3.182$ and calculate the area:

$$A = 4/3(20 - 3.182 - 2)^2 \left[\frac{(2)(30 - 3.182)}{(20 - 3.182 - 2) - 0.608} \right]^{1/2}$$

$$= 508 \text{ in}^2$$

Sample problem 5.5. Number of stays.

For the boiler discussed in sample problem 5.4, find the number of angle braces required. The angle braces are SA-675 grade 60, and they are 7/8-inch diameter. The ratio $L/1$ does not exceed 1.15. They will be welded per PW-19.

Solution: In this example, the number of braces will be determined by the strength of the braces. We begin by writing the basic stay equation and then evaluating each of the terms:

$$PA = naS$$

where P = MAWP = 200 psig; A = area to be stayed = pitch area – stay area; n = number of stays or braces; a = net area of a stay (no tell-tale holes in diagonal stays, and no threads); and S = allowable stress from Section II, Part D, Table 1A.

We note that PG-49.1 requires that a staybolt have its area multiplied by 1.1, equivalent to dividing the allowable stress from the table by 1.1. Further, PFT-28.1 states that the required area of staybolts and stays shall be as given in PG-49. However, PFT-32.2 indicates that for diagonal stays with $L/1 < 1.15$, the stays may be calculated as direct stays using 90 percent of the stress given in Table 1A. It is customary to apply only one factor in calculating stays—to use 1.1 on staybolts and direct stays and to use 90 percent of the tabulated allowable stress on diagonal stays meeting the conditions of PG-32.2. We therefore calculate the number of stays as shown below:

$$(200)\left[508 - n\left(\frac{\pi}{4}\right)(7/8)^2\right] = n\left(\frac{\pi}{4}\right)(7/8)^2(15{,}000)(0.9)$$

$$n = 12.33 \text{ stays (use 13)}$$

Had we used an area of 519 in^2, we would have obtained a required number of stays as 12.60 (use 13).

Photo 5.3 Diagonal stays in a Scotch firetube boiler. The view is of the head, from inside the shell during retubing. Photo by the authors.

Sample problem 5.6. Thickness of stayed plate.

For the boiler of sample problems 5.4 and 5.5, find the required plate thickness.

Solution: For SA-285 grade C, the allowable stress from Section II, Part D, Table 1A is 13,800 psi. In the absence of other information, we assume a square pitch. The total area is 508 in^2, and the number of stays is 13, which means that each stay carries 508/13 = 39.08 in^2. The pitch will therefore be the square root of 39.08, or $p = 6.25$ inches. We then go to PG-46 to compute the required plate thickness:

$$t = p\left(\frac{P}{SC}\right)^{1/2}$$

where t = minimum required thickness; p = pitch of stays; P = MAWP = 200 psig; S = allowable stress from Table 1A; and C = constant from PG-46.1 = 2.2 for welded stays. We get:

$$t = 6.25\left[\frac{200}{(13{,}800)(2.2)}\right]^{1/2}$$

$$= 0.507'' \text{ (use } ^9\!/_{16})$$

In laying out the tubes to support the head, we note that the greatest distance from the shell to the first row of braces is given by PFT-27.2 as the inside radius of the flange plus the pitch from PG-46. In this case, that would be 3 inches – $^9\!/_{16}$ + 6.25 = 8.69 inches (see Figure A-8 i and j).

Sample problem 5.7. Size of diagonal stays (photo 5.3).

The area to be stayed is 648 in^2, and the pressure is 150 psig. There will be 18 diagonal stays made of SA-36 material. The slant length of the stays will be 60 inches, and the straight distance along the shell from the head to the point of attachment is 50 inches. Find the minimum required diameter of the stays.

Solution: Each stay will carry 648/18 = 36 in^2 of plate. Therefore, the pitch will be $(36)^{1/2}$ = 6 inches. We write the basic stay equation:

$$PA = naS$$

where P = MAWP = 150 psig; A = area subjected to pressure = 648 – $18a$; a = area of stay; n = number of stays = 18; and S = allowable stress from Table 1A = 12,600.

We note that we can write the equation in terms of one or eighteen stays:

$$150(36 - a) = 12{,}600a$$

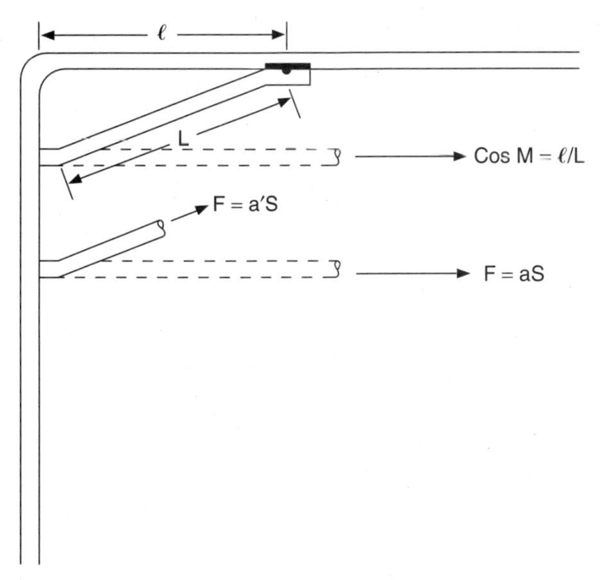

Sketch 5.4 $F = F' \text{ Cos } M = F'\ell/L$; $aS = a'S\ell/L = a' \text{ Cos } M$; $a' = a/\text{Cos } M = aL/\ell$ (PFT-32.1). When $M \le 30$ and $L/\ell \le 1.15$, use 1.15 per PFT-32.2.

or

$$150(648 - 18a) = 18(12{,}600)a$$

$$a = 0.427 \text{ in}^2$$

Then per PFT-32.1, the area of the stay must be multiplied by the ratio L/ℓ to arrive at the required area:

$$0.427(60/50) = 0.512 \text{ in}^2$$

The diameter can now be computed:

$$a = \frac{\pi d^2}{4}$$

$$0.512 = \frac{\pi d^2}{4}$$

$$d = 0.808'' \text{ (use } ^{13}\!/_{16})$$

We note that the ratio of the lengths (ℓ/L) is the cosine of the angle made by the stay with the axis of the boiler (sketch 5.4). The force in the diagonal stay is equal to the force normal to the head divided by the cosine of the angle. Therefore the area of the stay is increased in this

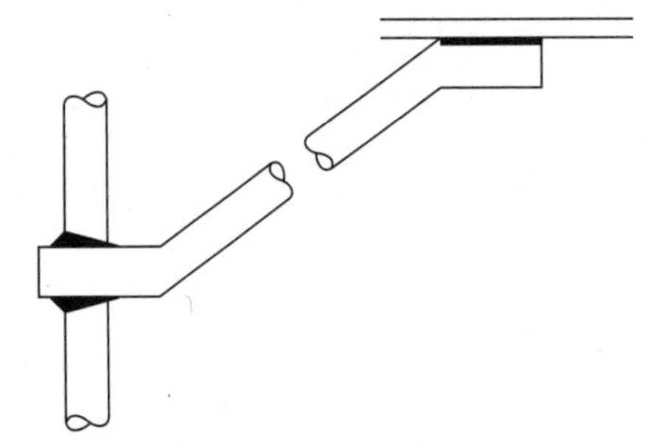

Sketch 5.5 Welding of diagonal stays (PW-19). Fillets ⅜″ minimum, full length both sides. Thick black areas indicate fillets.

proportion over the area required for a straight stay. Thus $F \cos M = PA$; $F = PA/\cos M$; $a = PA/S \cos M = PAL/S\ell$, where F = force in the diagonal stay; M = the angle between the diagonal stay and the centerline of boiler shell; P = MAWP; A = net area under pressure = pitch area minus stay area; a = stay area; and S = allowable stress in stay from Table 1A.

Sample problem 5.8. Weld size of diagonal stays.

Find the fillet weld size and the length of the welds attaching the diagonal stays of sample problems 5.5 through 5.7. Find also the weld requirements for the weld attaching the diagonal stay to the head.

Solution: PW-19.1 states that the weld attaching the stay to the head shall be a full penetration weld, and that the weld area in shear shall

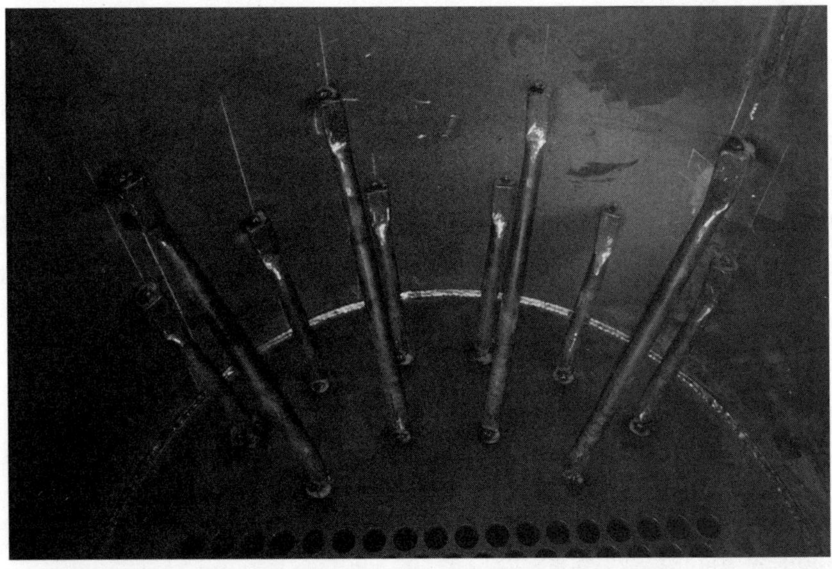

Photo 5.4 Welded diagonal stays for the head of a firetube boiler are shown tack-welded in place. Photo by the authors.

be at least 1.25 times the required stay area. Further, the weld size must be at least ⅜ inch. The required stay area is 0.512 in^2, and the stay diameter will be $^{13}/_{16}$ inch. The head thickness is $^{9}/_{16}$ inch. The weld area in shear is the product of the weld circumference and the sum of the plate thickness and the fillet weld leg:

$$1.25(0.512) = \pi(^{13}/_{16})(^{9}/_{16} + \ell)$$

$$\ell = -0.31$$

Therefore the weld at the head will be adequate if a full penetration weld with a fillet weld is used. The maximum projection of the ends of the stays beyond the head is ⅜ inch if the ends are exposed to combustion products. Any fillet weld size up to ⅜ inch is therefore permissible.

The welds joining the diagonal stays to the shell must be fillet welds (PW-19.4, Figure PW-19.4, and sketch 5.5). The size of the fillet welds must be at least ⅜ inch, and the throat area of the welds must total at least 1.25 times the required stay area. Welding across the end of the stay is optional (see photo 5.4), but such welds are not counted in the weld area computation.

The required stay area is 0.512. The required fillet weld throat area is therefore 1.25(0.512), and this is equated to the weld throat times the weld length:

$$1.25(0.512) = 0.7(^{3}/_{8})(L)$$

where the weld throat equals the leg times the cosine of 45°, which is customarily rounded to 0.7 in the Code, and where L = total fillet weld length.

$$L = 2.44''$$

The fillet welds could therefore be ⅜ inch in size and 1¼ inches in length on each side of the portion of the stay in contact with the shell. It should be noted, however, that the fillet welds must extend the full length of the portion of the diagonal stay that is in contact with the shell. Therefore, if it proved desirable to bend the ends of $^{13}/_{16}$-inch stays such that more than 1¼ inches contacted the shell, the weld would have to be increased in length also. In general one would expect the length of the stay in contact with the shell to be about 2½ to 3 diameters.

Sample problem 5.9. Area to be stayed of a head having a flanged-in manhole.

A horizontal return tube boiler is 72 inches in diameter and has a flanged-in manhole in the segment to be stayed in the area below the tubes. The distance from the lower row of tubes to the shell is 18 inches. The head thickness is $\frac{9}{16}$ inch, and the material is SA-516 grade 60. The outer radius of the flange is 3 inches. Find the area of the head that requires staying. The MAWP is 150 psig.

Solution: PFT-24.1 is used to find the value of d, which is either the outer radius of the flange but not more than 8 times the shell thickness (d = 3 inches but not more than $8(\frac{9}{16})$ or 4.5 inches) or $80t/(P)^{1/2}$ ($d = 80(\frac{9}{16})/(150)^{1/2} = 3.67$ inches). We may choose the larger value, and this will result in a lower area to be stayed, or we may choose to use 3 inches for simplicity, since this would allow us to use Table A-4.

Table A-4 gives 476 in^2 to be stayed, and this may be reduced by 100 in^2 because of the flanged-in manhole (PFT-27.9), provided the requirements of PFT-27.9.1 and PFT-29.1.2 are met. The area would therefore be 376 in^2.

We can choose to calculate the area per PFT-24.3.1:

$$A = 4/3(H - d - 2)^2 \left[\frac{2(R - d)}{(H - d - 2) - 0.608} \right]^{1/2}$$

$$= 4/3(18 - 3.67 - 2)^2 \left[\frac{2(36 - 3.67)}{(18 - 3.67 - 2) - 0.608} \right]^{1/2}$$

$$= 436 \text{ in}^2$$

This value can then be reduced by 100 in^2, as above. The answer is therefore $A = 336 \text{ in}^2$. The area below the tubes must be supported by through stays per PFT-24.4, which also requires that the clear distance between the through stays be at least 10 inches. This latter requirement is to permit inspection of the shell.

Sample problem 5.10. Maximum pitch of staybolts on a vertical firetube boiler furnace.

A vertical firetube boiler has a 40-inch diameter furnace made of $\frac{1}{2}$-inch plate with an allowable stress of 15,000 psi. Find the maximum pitch of the staybolts if the maximum allowable working pressure is to be 160 psig.

Solution: We note that the Code contains a strong inference that furnaces of this type over 38 inches diameter require staying as a flat plate. By reading PFT-23.3.1 and PFT-23.3.2, one can infer that furnaces over 38 inches *will* require staying, while furnaces 38 inches and under *may* require staying or may be plain circular furnaces as per PFT-14. See also the rule in PFT-23.5 for conical furnace tops. In this case, with the diameter over 38 inches, PFT-23.3.2 states that the maximum pitch will not exceed 1.05 times the pitch calculated by the rules of PG-46:

$$p = \frac{t}{(P/SC)^{1/2}}$$

$$= \frac{0.5}{160/(15{,}000)(2.2)^{1/2}}$$

$$= 7.18''$$

Therefore the maximum pitch for the 40-inch vertical firetube boiler will be 1.05(7.18) = 7.54 inches.

Sample problem 5.11. MAWP on a stayed box header.

A stayed flat box header is formed by welding. The header is 5.75 inches in the flat or trough. The thickness is ½ inch, and the allowable stress is 13,800 psi. Find the MAWP for the header.

Solution: PWT-12.1 gives the design rule for the welded stayed header, illustrated in Figure PWT-12.1. The flat trough may not exceed 90 percent of the pitch for the plate on the sides of the box header. If one assumes that the staybolts will be either welded or screwed, then for ½-inch plate the value of C will be 2.2. Therefore the pitch is equal to 5.75/0.9 = 6.39 inches.

PG-46.1 gives:

$$P = \frac{t^2 SC}{p^2}$$

where P = MAWP; t = plate thickness = ½ inch; S = allowable stress from Section II, Part D, Table 1A = 13,800 psi; C = 2.2 from PG-46.1 for welded or screwed stays where the plate is over 7⁄16 inch thick; and p = the pitch of the stays in the box header = 6.39 inches.

Therefore:

$$P = \frac{(0.5)^2(13{,}800)(2.2)}{6.39^2}$$

$$= 186 \text{ psig}$$

We note that the box header must have its weld radiographed and postweld heat-treated. We further note that the rule for welded, stayed waterlegs is given in PWT-12.2 and illustrated in Figure PWT-12.2. It should be noted that the waterleg joint does not require radiography.

Sample problem 5.12. Pressure on the conical top of a wet-top vertical firetube boiler.

A submerged top vertical firetube boiler has a conical top made of plate with an allowable stress of 15,000 psi. The thickness is ½ inch. The outside diameter at the bottom of the cone is 36 inches. The diameter at the top of the cone is 18 inches. The conical angle is 45°. Find the MAWP on the conical top, assuming it is unstayed.

Solution: PFT-23.4 indicates that such a submerged conical top may be used if the pressure (or thickness) is computed as if the cone were a cylinder, the outside diameter of which is the same as the greatest OD of the cone. We must therefore find the MAWP on a ½-inch-thick cylinder 36 inches in outside diameter. The Code does not indicate whether the length of the cylinder should be the slant length of the cone or the altitude. However, PFT-23.5 deals with conical tops of submerged vertical furnaces over 38 inches OD at the bottom (which require staying). This paragraph indicates that when calculating the unstayed portion of the conical top, the distance L should be the vertical distance or altitude of the cone rather than the slant length. It would seem clear that the same practice should be followed for unstayed cones. The MAWP will therefore be calculated using PFT-51, with $D_o = 36$ inches, $L = 9$ inches, $t = ½$ inch, and the temperature assumed to be 500°F. $L/D_o = 9/36 = 0.25$. $D_o/t = 36/0.5 = 72$. From Section II, Part D, Figure G, Factor $A = 0.012$. Steel with an allowable stress of 15,000 psi will require the use of Chart CS-2, where Factor B is found to be approximately 15,000. Then PFT-51.1.2, step 6 gives:

$$Pa = \frac{4B}{3(D_o/t)}$$

$$= \frac{4(15{,}000)}{3(72)} = 278 \text{ psig}$$

We then check to see if the temperature assumption is a good one. The absolute pressure is 278 + 15 = 293 psia. The steam tables give the saturation temperature for steam at 293 psia as 415°F. The design temperature for the furnace must be 100°F higher than the water temperature (PFT-17.7). Therefore, the design temperature must be 515°F. We further note that the absolute pressure for a saturation temperature of 400°F is 245 psia or 230 psig. The boiler could be rated for a maximum allowable working pressure of 230 psig based on a temperature of (400 + 100), or interpolation could be used between the 700° curve data and the 500° curve data to arrive at a value of Factor B for 515°F. This would permit a somewhat higher MAWP, above 230 but less than 278 psig.

We note that if the conical top is over 38 inches in outside diameter, it must be "fully supported" by staybolts (PFT-23.5). The pitch of the staybolts may be 1.05 times the pitch from PG-46. "Fully supported" means

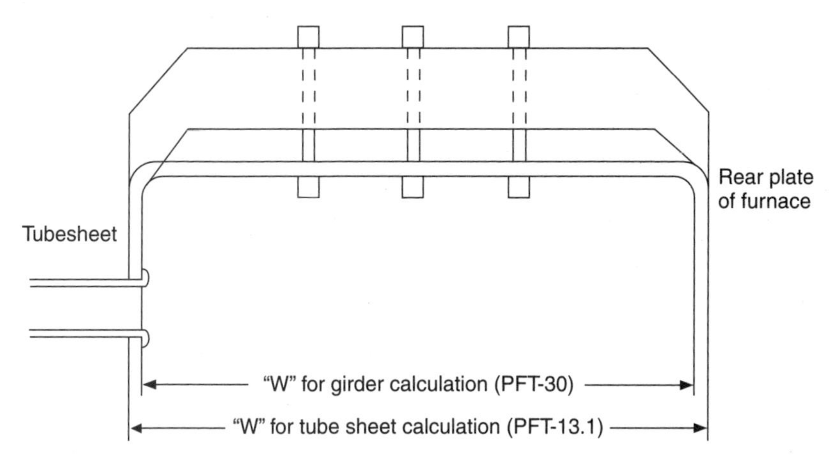

Sketch 5.6 Girder stays and crown bars.

that no credit is taken for the curvature (other than the 5 percent permissible increase in pitch). The plate is considered to be flat. The portion of the cone that must be stayed runs (at least) from the bottom up to the height, where the outside diameter is 30 inches. The pitch of the stays is measured on the inclined surface of the cone, and the stays are generally normal to this surface. We also note that the unstayed top portion

Photo 5.5 A sling stay and girder stay from a locomotive boiler. The shell supported the top of the furnace by means of the sling stay and girder stays. Photo by the authors.

of the cone must be calculated to ascertain whether it is capable of carrying a pressure at least as great as that carried by the stayed lower portion (or at least as great as the MAWP of the boiler).

Sample problem 5.13. Girder stays.

The top plate of the combustion chamber of a Scotch marine boiler is supported by staybolts that do not extend through another plate. Rather, the staybolts are supported on top of a beam (the girder) that runs across the top of the combustion chamber. The ends of the girders are carried on the tubesheet and the rear firebox plate (see sketch 5.6). The girders are made up of pairs of plates, with the staybolts extending down between them. The center-to-center distance of the bolts along the girder is generally the same as the center-to-center distance between girders across the boiler. This provides a square staybolt pitch on the top plate of the combustion chamber, which is the plate to be supported (see photo 5.5). Girders are also sometimes called crown bars. The plate must be designed based on the allowable stress, the MAWP, and the staybolt pitch. The staybolts must be sized based on the material allowable stress, the MAWP, and the pitch area. The girders must also be designed in accordance with the formula given in PFT-30.1. The plate and staybolt calculations have been illustrated earlier.

Find the MAWP on the combustion chamber of a Scotch marine wetback boiler if the top of the combustion chamber is supported by girder stays. The distance from the tubesheet to the back connection plate is 40 inches. The girder is 7 inches deep and ⅝ inch thick (for each plate, so that the total thickness is 1¼ inches). The girder pitch and bolt pitch are each 8 inches.

Solution: PFT-30.1 gives:

$$P = \frac{Cd^2t}{(W - P)D_1W}$$

where P = MAWP; C = a constant with a value that depends on the number of staybolts on each girder (in this case there are four bolts on each girder, so C = 11,000); d = girder depth = 7 inches; t = girder thickness = 1.25 inches; W = combustion chamber distance from the inside of the tubesheet to the inside of the back connection plate; p = staybolt pitch = 8 inches; and D_1 = girder pitch = 8 inches.

$$P = \frac{(11{,}000)(7)^2(1.25)}{(40 - 8)(8)(40)}$$

$$= 66 \text{ psig}$$

We emphasize that this pressure is based on the bending strength of the girder stays. The allowable pressure may be increased by increasing the depth and/or thickness of the girders or by reducing the span W. Increasing the girder depth to 8 inches and the girder plate thickness to ¾ inch (1.5 inches total) increases the MAWP to 103 psig.

Sample problem 5.14. Maximum allowable working pressure on a tubesheet in a combustion chamber.

Find the thickness required for a tubesheet in a Scotch marine boiler if the crown sheet is not suspended from the shell and the MAWP is 150 psig. The least horizontal pitch of the tubes is 4.25 inches and the tube inside diameter is 2.870 inches. The distance from the tubesheet to the back connection plate is 36 inches.

Solution: PFT-9.2.1 gives a table of minimum tubesheet thicknesses as a function of boiler shell diameter. The required thickness for a tubesheet, however determined, may in no case be less than that shown in PFT-9.2.1. However, when the crown sheet of a firebox is stayed by use of bolts carried by girder stays, the ends of the girders must be supported on the tubesheet on one end and on the back connection plate on the other end. The tubesheet is supported against the pressure by the tubes, and by stays if any are needed, but the plate is weakened by the tube holes in its ability to carry the point loads applied by the ends of the girder stays.

PFT-13.1 gives the formula for a combustion chamber tubesheet that is carrying the ends of the girders:

$$P = \frac{27{,}000t(D - d)}{WD}$$

where P = MAWP = 150 psig; t = tubesheet thickness; D = the least horizontal tube pitch = 4.25 inches; d = the inside diameter = 2.870 inches; and W = the distance from the tubesheet to the back connection plate, measured from the outside of the tubesheet to the outside of the opposite combustion chamber plate (the back plate), per the example in PFT-13.1. W = 36 inches. (Note that in sample problem 5.13, by PFT-30.1 the term W was the distance from the tubesheet to the opposite sheet measured *inside* the combustion chamber.)

$$t = \frac{PWD}{(27{,}000)(D - d)}$$

$$= \frac{150(36)(4.25)}{27{,}000(4.25 - 2.87)}$$

$$= 0.616'' \text{ (use } 5/8'')$$

This result is acceptable per PFT-9.2.1.

Chapter

6

Reinforcement of Openings in Heads and Shells

When openings are cut into heads or shells for nozzle openings, hand holes, manholes, and the like, the strength of the component is reduced. If the openings are small enough and far enough apart, the effect on the strength of the vessel may be negligible. As we saw in chapter 2, the method of compensating for a large number of openings in a close pattern is to increase the thickness of the plate so that the total plate cross-sectional area along a line through the center lines of holes in the pattern is as large as the area of an otherwise similar (but thinner) plate having no holes (PG-52 and 53). As we saw in chapter 3, openings may be reinforced by flanging them in and increasing the thickness (PG-29.3 and PG-34). Sometimes the shell or head has extra thickness, beyond the required minimum thickness for the MAWP, and this extra thickness may be sufficient to provide the needed strength for an opening. Sometimes reinforcement must be added in the form of a ring of steel around the opening, welded to the vessel.

Additional rules for openings and reinforcement (also called compensation) for cases not previously discussed are contained in PG-32, 33, 34, 35, 36, 37, 38, 39, 43, and PW-14, 15, and 16.

Very Small Openings in Shells, Headers, and Formed Heads

Welded connections up to 2 inches nominal pipe size (abbreviated NPS) may be attached to shells, headers, or formed heads, provided they are attached by "applicable rules" (PG-32.1.3.1.1). No calculations are required. PG-39.2 states that the applicable rules are given in PW-15 and PW-16. The rules for small welded connections are illustrated in

Figure PW-16.1(u), (v), and (w). PW-16.4 and 16.5 exempt welded connections below 3 inches NPS from complying with the minimum weld sizes shown in Figure PW-16.1 but apply other minimal requirements. PG-39.7 requires postweld heat treatment.

No calculations are required for threaded, studded, or expanded connections if the hole in the vessel wall does not exceed NPS 2 (PG-32.1.3.1.2). In the Appendix, A-65 discusses the requirements for a threaded 2-inch NPS connection to a boiler shell.

The rules for studded connections are given in PG-39.4.

The rules for threaded connections are given in PG-39.5. Table PG-39 shows the requirements for the number of threads and minimum plate thickness for the threaded hole based on the pipe size and pressure. The maximum pressure for threaded connections for boiler external piping is tabulated in PG-39.5.2, as a function of pipe size. The maximum pipe size for threaded connections to the boiler for boiler external piping is 3 inches NPS, and the corresponding maximum pressure is 400 psig. Smaller pipes may be used at higher pressures per the table. Threaded joints may not be used as boiler connections at temperatures above 925°F.

The rules for expanded connections are given in PG-39.6. Expanded connections may not exceed 6 inches OD. Blowoff piping for firetube boilers is covered by PFT-49, however, and limitations on the types of attachments are found therein: If the blowoff piping is exposed to the products of combustion, the attachment to the boiler must be by a threaded connection. If the blowoff pipe is not exposed to the products of combustion, it may be attached by any method except expanding.

Small to Medium Openings in Cylinders and Formed Heads Over 2 inches NPS to 8 inches OD

When the size of the opening exceeds 2 inches NPS but does not exceed 8 inches outside diameter, the need for compensation may be checked on Figure PG-32. This chart is entered at the bottom axis, using the value of the product of the drum diameter times the drum thickness. At the value of Dt, one moves vertically on the chart to the intersection of the curved line having the value of the term K (the computation of which is given below). From the intersection of Dt and K, one moves to the left axis and reads the value d of the largest unreinforced opening that may be installed in the subject drum. The largest opening permitted by this chart is 8 inches. It should be noted that d is the "maximum allowable diameter of opening" in the shell when the method of attachment is expanding, threading, or studded connection, because in these methods the nozzle wall offers no reinforcement to the shell. When the attachment is by welding in accordance with PW-15 and 16, d is the inside diameter of the nozzle (see Figures PG-36 and PW-16.1). The value of K is given in PG-32.1.2:

$$K = \frac{PD}{1.82St}$$

where P = MAWP; D = OD of the shell; S = allowable stress in shell material from Table 1A of Section II, Part D; and t = shell thickness.

The value of d may be read directly from the chart in Figure PG-32, or it may be calculated with the equation given in Figure PG-32:

$$d = 2.75[Dt(1 - K)]^{1/3}$$

The equation may not be used to extend the range of the chart past d = 8 inches or past Dt = 600. It should also be noted that the largest value of K that may be used in this equation is 0.990.

The use of Figure PG-32 is governed by rules in PG-32.

For shells and headers, PG-32.1.3.2 states that no calculation is required to determine compliance with the reinforcement requirements of PG-33 if the diameter of the opening does not exceed the value of d found by referencing Figure PG-32. Thus, for shells and headers, no reinforcement calculations are needed for openings 2 inches NPS or smaller, or for those smaller than d from the chart, provided that they are "single openings." The Code is silent as to how far apart two such openings must be to be considered "single."

For torispherical heads, no compensation calculations need be made if the opening is a single opening, if the opening is properly located on the head, and if the diameter of the opening does not exceed 2 inches NPS or that permitted by Figure PG-32.

PG-32.1.4.1 gives a formula for determining the minimum center-to-center distance between two openings on a formed head if the openings are to be considered "single":

$$L = \frac{A + B}{2(1 - K)}$$

where L is the minimum center-to-center distance between the openings, measured along the surface; A and B are the outside diameters of the two openings; $K = PD/1.82St$; P = MAWP; D = OD of the head; S = allowable stress for the head material; and t = head thickness.

PG-32.1.4.2 requires that the opening be located at least one head thickness away from the knuckle and from any flanged-in manhole. The opening must be in the spherical portion of the torispherical head. (The only openings permitted in the knuckle radius area of a torispherical head are the water column connections. These are required by PG-60.3.4 to be at least 1 inch, and generally would not be expected to be larger.)

PG-32.1.4.3 indicates that the maximum opening permitted in a dished head must not be larger than would be permitted in an "equivalent shell." The equivalent shell would have the same allowable stress, outside diameter, and MAWP as the head. It would, however, have a different thickness, generally smaller. But PG-32.1.5 states that no calculation need be made to demonstrate compliance when the diameter of a single opening does not exceed the diameter permitted by Figure PG-32. Therefore it is clear that the intent of the Code is to limit the placement of the openings to the spherical portion of the dished head, to require a minimum distance between adjacent openings, and to allow 2 inches NPS and smaller openings and openings not over d from PG-32 without requiring compensation calculations.

Semiellipsoidal heads are treated similarly to standard dished (torispherical) heads. The minimum center-to-center distance is given by PG-32.1.4.1, as above. PG-32.1.4.2 requires that the opening be at least one head thickness away from a flanged-in manhole or the edge of the ellipsoidal (central) area of the head. PG-32.1.4.3 again refers to the "equivalent shell" (which in the case of an ellipsoidal head would have the same required thickness as the head), and PG-32.1.5 again authorizes the use of Figure PG-32. Therefore, for semiellipsoidal heads, single openings do not require compensation calculations if they are not larger than 2 inches NPS or not larger than permitted by the chart, as long as they are not within one head thickness of the flanged-in openings or the edge of the rounded part of the head.

Hemispherical heads are treated similarly to other formed heads, but the value used for K in computing the minimum distance between single openings (PG-32.1.4.1) and in using the chart in Figure PG-32 is one-half the value computed per PG-32.1.2. (Note that the term K is defined in PG-32.1.2 and again in PG-32.1.4.1. Both definitions are the same.) Thus, single openings in hemispherical heads do not require reinforcement calculations if they do not exceed 2 inches NPS or the value of d from Figure PG-32, found by using $K/2$ rather than K from PH-32.1.2. The openings are "single" if the center-to-center distance is at least L, as computed in PG-32.1.4.1, again using $K/2$, where K is defined in PG-32.1.2.

An example of the use of Figure PG-32 is given in A-68 of the Appendix.

Compensation for Openings in Cylinders and Formed Heads

When the size of the opening in a shell or formed head exceeds 8 inches, or when it exceeds 2 inches NPS and the value of d found in Figure PG-32, the opening requires reinforcement calculations as described in

PG-33, 36, 37, 38, and PW-15 and 16. In short, when an opening does not meet the requirements for special exemption, compensation calculations must be performed. The basis of the calculations is explained below, and sample problems follow.

As discussed in chapter 2, internal pressure causes a state of biaxial stress in a cylinder wall. The circumferential stress on the cylinder wall is the largest principle stress, and the longitudinal stress is one-half as great. In sketch 6.1, a hole has been cut in the wall of the vessel. The principal stress direction is represented in the sketch. As the vessel wall resists the pressure force, the stresses adjacent to the hole are higher than the average stress remote from the opening. Thus, the summation of the stress-area products is still equal to the pressure force carried by the cylinder wall. If, however, the pressure was the MAWP for the thickness of the original shell, then the average stresses were already at the maximum allowable stress value for the material and temperature. Cutting the hole will cause a local overstress condition, as the material near the hole carries the force that would otherwise be carried by the missing metal (see sketch 6.2). The magnitude of the extra force to be carried is equal to the product of the hole diameter times the required thickness for the shell times the allowable stress for the material. This extra force must be carried by the material on both sides of the hole and within a limited distance to either side of the hole. Because the shell material is already at its minimum required thickness, it is already stressed to the maximum allowable stress. Extra metal must be provided or found within a relatively short distance of the hole, on either side, and above and below the surfaces of the shell. The distances above and below the shell surfaces and to the left and right of the opening in the shell within which extra metal may be considered to carry the extra stress are called the limits of compensation. The extra metal inside the limits of compensation can carry a force equal to the product of the metal area times the allowable stress. Therefore the compensation calculations consist of identifying extra metal area inside the limits of compensation such that the extra metal area times the allowable stress is at least as great as the product of the hole diameter times the required thickness times the allowable stress. When there is not sufficient extra metal area within the limits of compensation around an opening, a reinforcing pad may be installed on the outside of the shell, surrounding the hole. The pad may also be placed on the inside, or sometimes pads or rings will be used on both surfaces of the shell.

Sketches 6.3 and 6.4 illustrate the extra metal area in the shell wall that may be considered to reinforce the opening in the shell within the limits of reinforcement. Each area lies in the plane of the cut section. Tensile stresses equal in magnitude to the maximum allowable stress

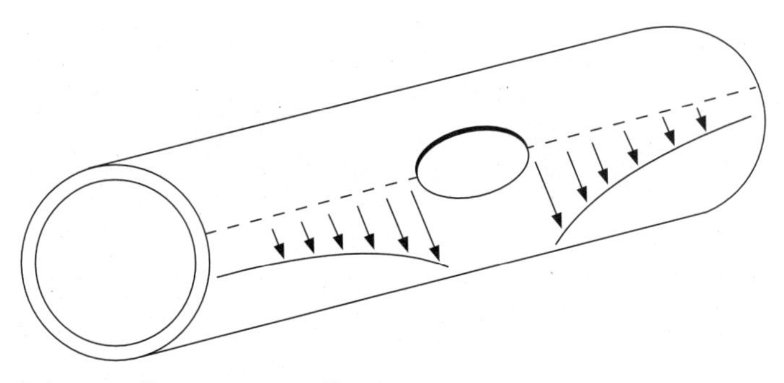

Sketch 6.1 Stress at an opening.

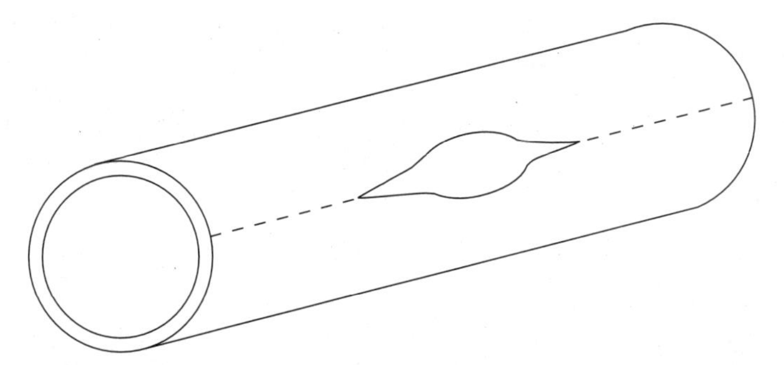

Sketch 6.2 Manner in which an opening would tear if it was not reinforced.

are considered to act on these extra areas, and the summation of the areas times their allowable stress must at least equal the hole diameter times the required thickness times the allowable stress for the shell. (For purposes of this discussion all areas are considered to have the same maximum allowable stress. If all areas have the same allowable stress, then areas are directly proportional to the strength and may be counted directly in computing compensation. Thus compensation is normally computed in square inches rather than pounds or psi). The areas within the limits of compensation that may be considered to have value as compensation are:

1. *Extra thickness of the shell (sketch 6.3).* If the shell is thicker than required for the MAWP, the extra thickness of the shell on each side of the opening can be counted. Note that for purposes of computing compensation, the shell is considered to extend all the way to the inside of the finished opening.

2. *Projection of the nozzle into the vessel.* The portion of the nozzle that projects into the vessel experiences no pressure differential from inside to outside. Therefore there are no tensile stresses induced by the

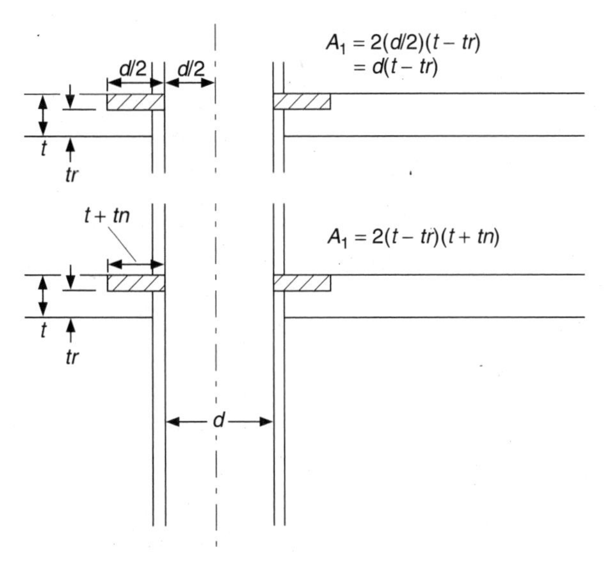

Sketch 6.3 Inherent compensation in shell uses the larger value.

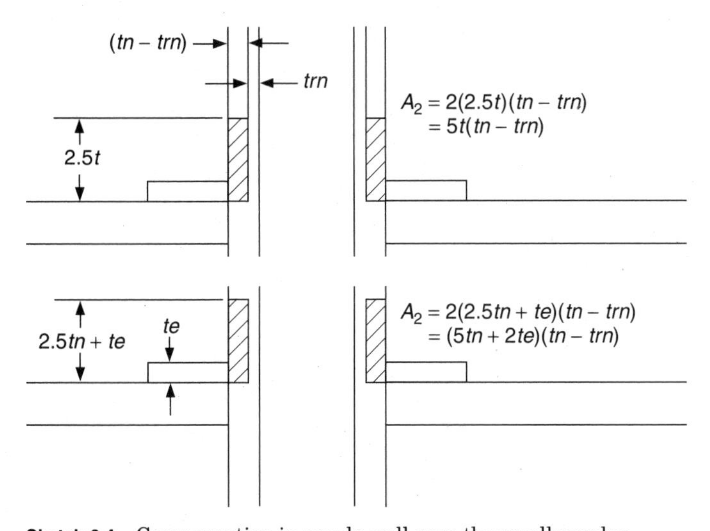

Sketch 6.4 Compensation in nozzle wall uses the smaller value.

pressure, and the full area projecting inwards from the inside of the shell can be considered as compensation.

3. *The fillet weld areas within the limits of compensation, all of which may be counted.* The groove welds are considered to be part of the compensation ring material, and will be so counted below.

4. *The portion of the nozzle that sticks out of the shell does contain pressure.* The thickness required to contain the pressure with a stress equal to the allowable stress is not available as compensation. The extra thickness of the nozzle, beyond what would be required for its internal pressure, is available on both sides of the opening out to the limit of compensation on the outside of the vessel shell (sketch 6.4). PG-43 requires that in a compensated opening the nozzle neck thickness must be at least equal to the smaller of: the thickness of the vessel part to which it is attached; or the thickness of standard weight pipe (schedule 40) for nozzles made of pipe; or the minimum thickness for 600 psig if the nozzle neck is a tube nipple. These requirements are in addition to and independent of the compensation calculations.

5. *The reinforcing ring itself is, of course, compensation.* Its full cross section is counted on each side of the hole, out to the limit of compensation.

The amount of compensation needed for an opening is given in PG-33.2:

$$A = dt_rF$$

where A = the area of compensation needed for the opening in the head or shell; d = the inside diameter of the finished opening, i.e., the ID of the nozzle; and t_r = the minimum required thickness of the shell or formed head.

If the opening is to be in a cylindrical shell, t_r is computed by use of the formula in PG-27.2.2.

For a standard dished (torispherical) head, if the opening and its limits of compensation are contained completely within the spherical portion of the head, t_r is the required thickness for a hemispherical head, the radius of which is equal to the radius of dish of the head, using PG-29.11 (see PG-32.2.1).

If the head is semiellipsoidal, and if the opening and its limits of compensation are contained within a circle centered on the center of the head and having a diameter equal to 80 percent of the shell ID, then t_r is calculated as the thickness of a hemispherical head, the radius of which is 90 percent of the shell ID (see PG-33.2.2).

If the head is hemispherical, t_r is calculated by use of the formulae in PG-29.11, whichever applies.

F is a factor that may be used when the plane under consideration has a tensile stress less than the maximum principle tensile stress. For example, an elliptical opening in a cylindrical shell could be placed with its longest dimension in the circumferential direction. Then the stresses on the plane containing the longest direction of the opening would be less than the stresses on the plane containing the lesser axis of the elliptical opening. The area required to reinforce that plane would then be $dt_r(0.5)$, if the opening was to be integrally reinforced (PG-32.2). $F = 1$ for all circular openings, for all openings in heads, and for all openings in cylindrical shells where compensation is not integral. (Compensation is not integral if a ring is welded on.)

The limits of compensation are given in PG-36. Only metal within these limits may be counted as having value for compensation. The first limits are measured from the axis of the nozzle parallel to the surface of the vessel. The distance is the same on either side of the center line of the opening, and the larger of two choices is taken as the limit per PG-36.2. The choices are the diameter of the finished opening or the sum of the radius of the finished opening plus the vessel thickness plus the nozzle thickness (see Figure PG-36). The distance measured perpendicular to the surface of the shell is also one of two choices, with the smaller value being chosen (PG-36.3). The choices are: 2.5 times the shell thickness; or 2.5 times the nozzle thickness plus the thickness of the ring if a ring is being added. In any compensation calculation, it is useful to sketch in and dimension the limits of compensation before attempting to compute the available areas.

The areas that may be counted as compensation are explained in PG-36.4. PG-36.4.1 gives two formulae for computing the area in the vessel wall above that required by PG-27.2.2. The first formula is $A_1 = (t - Ft_r)d$, and the second formula is $A_1 = 2(t - Ft_r)(t + t_n)$. As can be seen in sketch 6.3, the first formula gives the amount of area available in the shell if the limit of compensation is the diameter of the finished opening. The second formula gives the amount of area available in the shell if the limit of compensation is given by the radius of the finished opening plus the vessel wall plus the nozzle wall. Therefore, if the limits of compensation have been worked out and sketched in, it is unnecessary to compute both formulae in PG-36.4.1. One can select the formula that corresponds to the limit of compensation in the case under consideration.

Similarly, PG-36.4.2 gives two formulae for computing the amount of metal area available as excess thickness in the nozzle wall. The first formula, $A_2 = (t_n - t_{rn})(5t)$, corresponds to the limit of compensation taken at 2.5 times the vessel wall thickness (see sketch 6.4). The second formula, $A_2 = (t_n - t_{rn})(5t_n + 2t_e)$, corresponds to the limit of compensation taken at 2.5 times the nozzle thickness plus the pad thickness. Again, once the limits of compensation have been selected and sketched, it is only necessary to use the one formula from PG-36.4.2 that corresponds to the limit chosen.

The additional areas that may be counted are identified in PG-36.4.2 and 36.4.3. PG-36.5 refers the reader to example problems in computing compensation in the Appendix, A-65 to A-69.

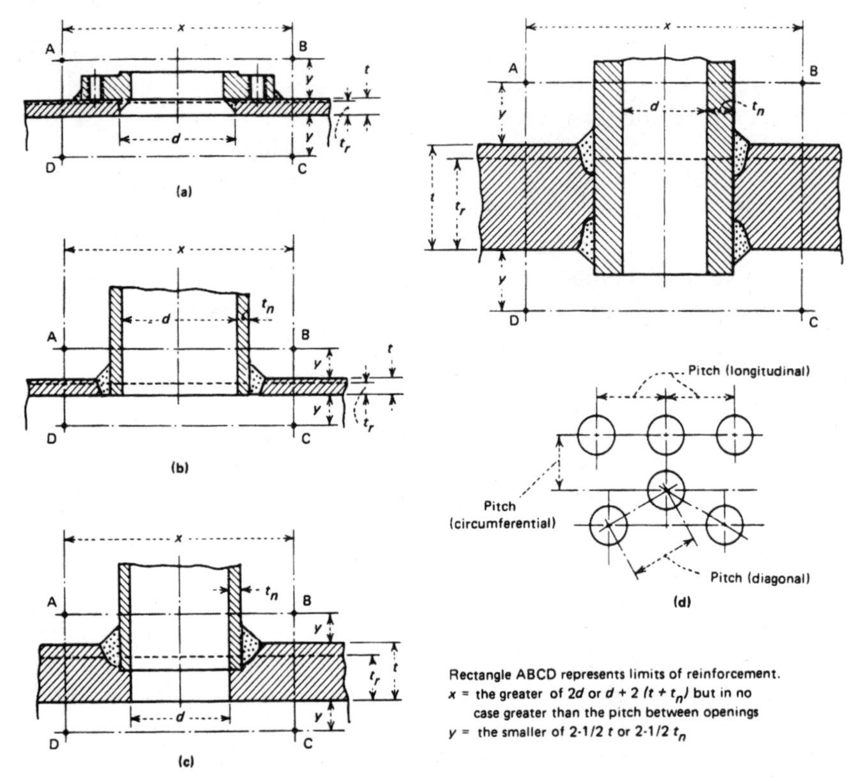

Figure PG-36 Limits of reinforcement for typical openings. Reprinted courtesy of ASME.

The rules in PG-33 and PG-36 describe the manner of computing the area of metal that may be considered as compensation. PG-37 is entitled "Strength of Compensation." PG-37.1 deals with the allowable stress values of the metal areas counted as compensation. In the simplest case, the shell, nozzle and pad (if any) all have the same allowable stress from Table 1A. In that event, all areas, including all welds, are considered to have an allowable stress equal to that of the shell. Then, if the area of compensation totals at least dt_rF, there is sufficient compensation.

If either the nozzle or the pad has a higher allowable stress than the shell, it is considered to have the same allowable stress as the shell for purposes of compensation. Again, the welds are considered to have the same allowable stress as the shell, and again the computation of the compensation is simply the totalling of the areas within the limits of compensation (see sketch 6.5).

If, however, the nozzle or the pad has a lower allowable stress than the shell, the area to be counted as compensation must be reduced by multiplying the actual area by the ratio of the allowable stresses. Fillet weld areas are considered to have an allowable stress value equal to that of the weaker of the parts being joined by the weld. Therefore, fillet weld areas must also be corrected downwards by multiplying by the appropriate allowable stress ratio. Groove welds joining the shell or pad to the nozzle are generally taken to have an allowable stress equal to that of the shell or pad. The areas, corrected as necessary, are then totalled to determine if there is sufficient compensation.

PG-37.2 deals with the strength of the attachment of the compensation and nozzle to the shell. A reinforcing pad would not be effective unless joined to the shell by welds of sufficient strength to transmit the load to be carried by the pad. The size of the load(s) to be carried through welds and the like on paths defined in PW-15 and Figure PW-16, are given in PG-37.2 as equations 1, 2, and 3. The first equation is:

$$W = dt_rS$$

where d is the diameter of the finished opening; t_r is the required thickness of the shell or head; S is the allowable stress of the shell or head material; and W is the load to be carried.

In equation 1, dt_r is the area that must be replaced to compensate the opening. Therefore, equation 1 represents the load that must be carried by the compensation in order to replace the missing metal.

Equation 2 is:

$$W = \{dlt_r - [(2d - dl)(t - t_r)] + A_S\}S$$

where dl is the diameter of the hole in the shell before nozzle installation; and A_S is the area of two stud holes assumed to lie in the plane of interest in the case where there are stud holes in the vessel wall. Otherwise, $A_S = 0$.

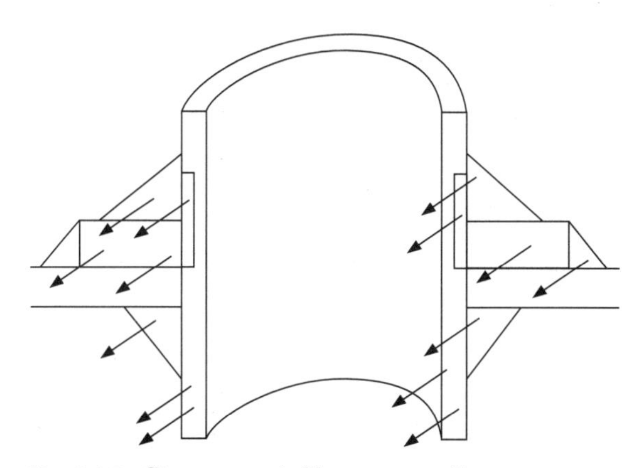

Sketch 6.5 Stresses carried by compensation.

In equation 2, dlt_r is the area of missing metal based on the unfinished opening, and $(2d - dl)(t - t_r)$ is the area of compensation available in the shell. Therefore, W in equation 2 is the difference between the total load based on the rough opening and the load carried by the compensation in the shell. This amount of load must be carried by the nozzle, fillet welds, and pad, if any.

Equation 3 is:

$$W = \{dlt_r - [2t(t - t_r)] + A_S\}S$$

where all terms are defined as for equations 1 and 2.

In equation 3, dlt_r is again the area of missing metal based on the unfinished opening. The area of the compensation in the shell based on the second of the two formulae shown in PG-36.4.1 is $2t(t - t_r)$. Therefore, W is again the difference between the total load to be carried (based on the unfinished opening) and the load carried by the extra thickness of the shell within the limit of compensation. W is the load that must be transferred through the attachment welds to be carried by the compensation in the nozzle wall and pads, if pads are needed.

As indicated above, it is necessary to compute the loads W from PG-37. The strength of each load path (to be discussed below and in PW-15 and Figure PW-16) must be at least equal to the smallest value of the three W's computed in PG-37. We note that W from equations 2 and 3 can be negative, and that negative values of W are taken as equal to zero. Therefore, in some cases, it may not be necessary to calculate the paths per PW-15 because the smallest value of W is zero. ($W = 0$ indicates that the area in the shell equals or exceeds the amount of area needed based on the unfinished opening. Therefore no compensation loads need to be carried by the nozzle wall and the attachment welds.) Even in cases where $W = 0$, however, it will still be necessary to verify that the size of the welds for welded attachments conforms to the requirements of PW-16. These requirements are specifically independent of the calculations discussed so far.

After computing the compensation areas (PG-33, 36, and 37.1), and after computing the three values of the load to be carried by the load-carrying paths (PG-37.2), it is recommended that the weld sizes be checked for compliance with PG-16. If the welds do not comply with the minimum size requirements of PG-16, the design does not meet the Code, even if the strength of the load-carrying paths is sufficient for the compensation load W. Further, it will generally be found that if the weld sizes do meet PW-16, the load paths will calculate up to strength.

There are a number of conditions where load path calculations are required, and other conditions where they are not required. These are summarized as follows:

1. The welded attachment is 2 inches NPS or smaller, and the welds meet the size requirements of PW-16. No further calculations are required.
2. The threaded, studded, or expanded opening is 2 inches NPS or smaller. There are no attachment welds, and no calculations are required.
3. The attachment is exempted from compensation calculations per Figure PG-32. If it is welded, the weld sizes must meet PW-16 requirements. Load path calculations per PG-37.2 and PW-15 are required as given in Interpretation I-95-02.
4. The welded attachment meets the requirements of PG-33 for availability of compensation. The weld sizes meet the requirements of PW-16. The least value of W, as computed per PG-37.2, is zero. No further calculations are required.
5. The welded attachment meets the requirements of PG-33 for availability of compensation. The weld sizes meet the requirements of PW-16. The least value of W calculated in accordance with PG-37.2 is greater than zero. The strength of the load-carrying paths must be calculated in accordance with PW-15, as discussed below.
6. The welded attachment is a fitting not larger than NPS 3 that is welded per Figure PW-16.1(u2), (v2), (w2), or (x). The attachment meets the requirements for compensation, either by the exemption for NPS 2 and smaller, the exemption per Figure PG-32, or by the requirements of PG-33. The weld size requirements of PW-16 need not be met, but the load path strength requirements of PW-15.1 must be met (see PW-16.4). This requires the load path calculations per PG-37.2 and PW-15.

The calculation of the load paths for compensated openings is one of the most misunderstood areas in the Boiler Code. First, it must be remembered that all of the compensation design is associated with preventing the vessel from tearing open along a plane passing through the opening, as illustrated in sketch 6.2 for the plane of the maximum principal stress in a cylindrical vessel fitted with a circular opening. Compensation carries the tensile stresses that would have been carried by the metal removed in cutting the hole, as shown in sketch 6.5. (For simplicity, in this sketch, $t = t_r$, so the vessel wall has no compensation capability. Also groove welds are not shown because they are taken as part of the shell or pad.) Because the tensile strength of the compensation has been designed per PG-33 and PG-36, the compensation ring will not fail in tension, nor will the nozzle tear into two half-cylinders. The attachment will not fail as long as the welds securing the compensation areas to the shell have enough strength to transmit the loads from the shell to the ring and nozzle. The forces acting to cause failure are the tensile forces represented in sketch 6.5. All paths of failure that must be considered would therefore involve a motion of the pad and nozzle as a whole; the welds attaching the pad and nozzle can fail, and the nozzle wall can shear off, but the compensation ring will not break, and the nozzle will not tear open longitudinally.

Figure PW-16 illustrates a number of types of welded attachments and indicates the load paths that must be calculated for each design shown. Not given, however, are the welds that must be considered to lie in the different paths. Accordingly, these are discussed below.

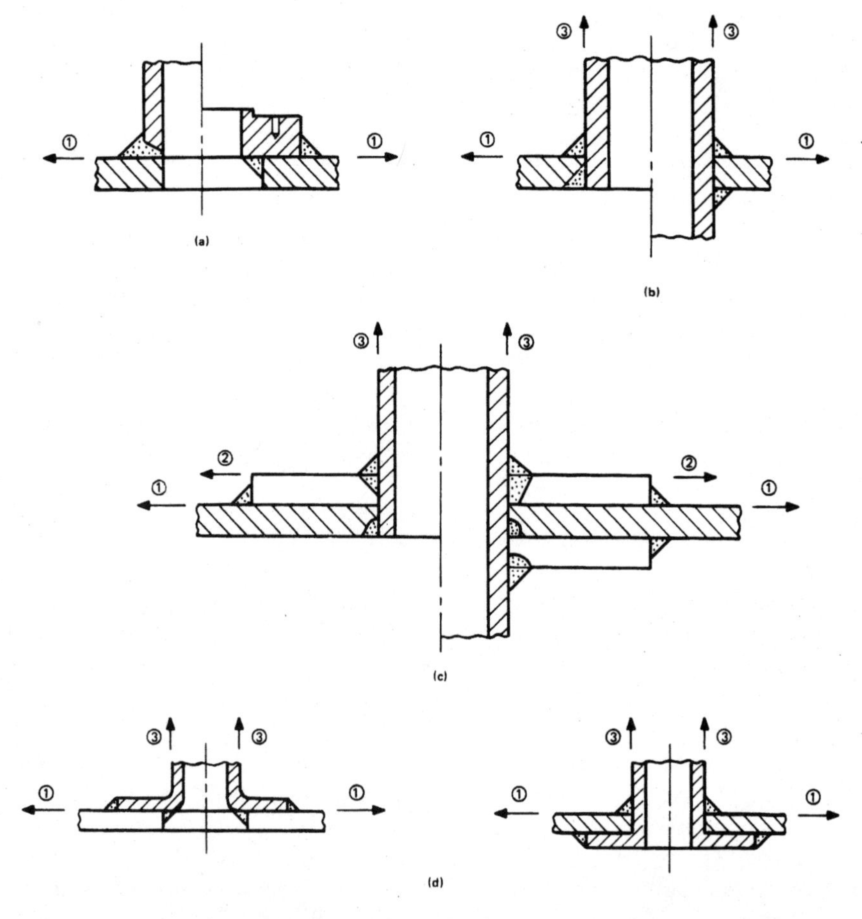

① Denotes the load-carrying path acting perpendicular to the nozzle center line about the nozzle at the face of the vessel.
② Denotes the load-carrying path acting perpendicular to the nozzle center line about the nozzle at the face of the external pad.
③ Denotes the load-carrying path about the nozzle acting parallel to the nozzle center line.

Figure PW-16 Load-carrying paths in welded nozzle attachments. Reprinted courtesy of ASME.

Figure PW-16(a), left side. Only path 1 is required, and path 1 consists of the fillet weld in shear plus the groove weld in shear. The areas in shear are measured on the surface of the vessel.

Figure PW-16(a), right side. Again, only path 1 is required. Both fillet welds are in shear, and the areas are measured in the plane of the surface of the vessel. (The curvature of the vessel surface is not considered in computing the areas of these welds.)

Figure PW-16(b), left side. Paths 1 and 3 are considered. Path 1 contains the fillet weld in shear and the nozzle neck in shear. Path 3 contains the fillet weld in shear and the groove weld in tension. The areas of both the fillet and groove weld are measured on the vertical surface of the nozzle. Contrary to common conception, however, path 3 does not represent pushing the nozzle out of the hole. Rather, it represents the shell tearing apart, such that the welds holding the nozzle to one side of the shell fail, allowing the nozzle to move with the other half of the shell. The groove weld is thus considered to be in tension. Fillet welds are always considered to be in shear.

Figure PW-16(b), right side. Again, paths 1 and 3 are required. Path 1 contains the nozzle and one fillet weld in shear, with its area being measured on the surface of the vessel. Path 3 contains two fillet welds in shear, with the areas being measured on the vertical surface of the nozzle.

Figure PW-16(c), left side. Paths 1, 2, and 3 must be considered. Path 1 consists of the fillet weld joining the pad to the shell and the nozzle wall in shear. The area of the fillet weld is measured on the surface of the vessel as the mean circumference times the leg size. The nozzle wall area is simply the cross-sectional area of the nozzle. Path 2 contains the pad-to-nozzle fillet weld in shear, with its area measured in the plane of the surface of the pad, and the nozzle wall, also in shear. Path 3 contains the pad-to-nozzle fillet weld in shear and the two groove welds in tension, with all areas being measured on the surface of the nozzle.

Figure PW-16(c), right side, paths 1, 2, and 3. Path 1 contains the pad-to-shell fillet weld and the nozzle wall in shear. Often, no credit is taken for the pad-to-nozzle groove weld. If, however, a definite minimum dimension is specified for the width of the weld to the shell, then that groove weld area may be counted in shear. Path 2 contains the pad-to-nozzle fillet weld and the nozzle wall. Path 3 contains the two pad-to-nozzle fillet welds in shear and the three groove welds in tension. All areas are measured on the surface of the nozzle.

Figure PW-16(d), left sketch, paths 1 and 3. Path 1 contains the two fillet welds in shear, with the areas to be measured in the plane of the surface of the vessel. Path 3 contains both the nozzle lip and fillet weld in shear.

Figure PW-16(d), right sketch, paths 1 and 3. Path 1 contains the fillet weld and the nozzle wall. Path 3 contains the fillet weld and the nozzle lip.

Sample Problems

Sample problem 6.1. Checking a small opening for inherent compensation per PG-32.

A 6-inch schedule 80 nozzle is fitted to a dished head. The head is 60 inches diameter and 1.5 inches thick. The material is SA-516, grade 70. The allowable stress on the nozzle material is 15 ksi. The maximum allowable working pressure is 400 psi. The nozzle projects 1.5 inches inside the inner surface of the head and is attached by fillet welds inside and out. Check the nozzle for adequacy of compensation and find the minimum weld size for the fillet welds. Find also the minimum distance between two such openings. Assume the nozzle is to be fitted with a flange and a valve. What ASME/ANSI class will be required for the flange and valve?

Solution: PG-32.1.4.1 gives:

$$K = \frac{PD}{1.82St}$$

where P = MAWP = 400 psig; D = OD of head = 60 inches; S = allowable stress of head material = 17,500 psi (Table 1A); and t = head thickness = 1.5 inches.

$$K = \frac{(400)(60)}{(1.82)(17{,}500)(1.5)} = 0.502$$

$$Dt = (60)(1.5) = 90.000$$

The equation from Figure PG-32 is:

$$\begin{aligned} d &= 2.75[Dt(1 - K)]^{1/3} \\ &= 2.75[90(1 - 0.502)]^{1/3} \\ &= 9.766 \end{aligned}$$

Therefore the maximum unreinforced opening in the subject head is 8 inches per Figure PG-32. Six-inch schedule 80 pipe has OD = 6.625 inches and thickness = 0.432 inch (nominal). If the thickness is reduced by 12½ percent, the minimum thickness will be 0.378 inch, and the ID of the opening will be 6.625 − 2(0.378) = 5.869 inches. The opening does not require reinforcement.

The welded attachment is as shown in Figure PW-16.1(d). The requirements are that the sum of the weld throats must be at least equal to 1.25 times t_{min}, and each weld must be at least equal to the smaller of t_{min} or 0.25 inch. PW-16.2 defines t_{min} as the smaller of ¾ inch or the thinner of the two parts being joined. In this case, t_{min} is equal to 0.378 inch, the nozzle thickness. Therefore $t_1 + t_2 = 1.25(0.378) = 0.473$. Each weld throat must be (0.7)(0.378) = 0.265 or 0.25 inch, whichever is less. Therefore $t_1 = t_2 = 0.25$. The weld leg will be 0.25/0.7 = 0.357 inch. A ⅜-inch fillet weld would be specified in this case.

The minimum distance between two such openings is given in PG-32.1.4.1:

$$L = \frac{A + B}{2(1 - K)}$$

where K is as defined above; A and B are the outside diameters of the adjacent openings; and L is the minimum center-to-center distance for the openings.

$$L = \frac{6.625 + 6.625}{(2)(1 - 0.502)}$$
$$= 13.303''$$

The flange and valve classes are given in Table A-361 of the appendix. The flange fitted to the nozzle must be ASME/ANSI B16.5 class 300 (good up to 605 psig), and the valve must be ASME/ANSI B16.34 class 300, again acceptable for use up to 605 psig.

Sample problem 6.2. Compensation for a threaded opening on a boiler.

A 48-inch HRT boiler is to have its 2½-inch NPS blowoff line screwed into a hole threaded directly into the $\frac{5}{16}$-inch-thick shell of the boiler. The boiler has an MAWP of 150 psig, and the allowable stress on the shell material is 13.8 ksi. Does the opening require added compensation? What class blowoff valve is required?

Solution: PG-32.1.2 gives:

$$K = \frac{PD}{1.82St}$$

where P = MAWP = 150 psig; D = shell diameter = 48 inches; S = shell allowable stress = 13,800 psi; and t = shell thickness = $\frac{5}{16}$ inch.

$$K = \frac{150(48)}{1.82(13{,}800)(\frac{5}{16})} = 0.917$$

$$Dt = 48(\tfrac{5}{16}) = 15.000$$

The equation in Figure PG-32 gives:

$$d = 2.75[Dt(1 - K)]^{1/3}$$
$$= 2.75[(15)(1 - 0.917)]^{1/3}$$
$$= 2.954''$$

The maximum diameter of the 2½-inch NPS threaded hole will be 2.875 inches. Therefore, no additional compensation will be needed. However, when PG-39 is checked for the adequacy of the threaded connection, it is found that 5⁄16 inch is too thin to provide the required seven engagement threads. A built-up pad or fitting may be used per PG-39.5.1 to provide the necessary 0.875 inch minimum thickness for threading. We note that PFT-49.1 requires that this attachment be made by threading because the blowoff pipe is exposed to the products of combustion. We also note that the blowoff valve must be at least class 150 in Table A-361, suitable for blowoff line service to 170 psig.

Sample problem 6.3. Large nozzle in a formed head.

A 16-inch nozzle is to be installed in a 72-inch semiellipsoidal head having a maximum allowable working pressure of 650 psig, with pressure on the concave side. The nozzle is centered on the head and is made of 16-inch XS pipe with an allowable stress of 15,000 psi. The head is 1.5 inches thick and has an allowable stress of 17,500 psi. The pad will be made of plate with an allowable stress of 13,800 psi. Design the opening, showing all necessary calculations.

Solution: First the required area of reinforcement is found. PG-33.2.2 indicates that t_r is the minimum required thickness of a seamless hemispherical head having a radius equal to 90 percent of the head inside diameter. PG-29.11, formula 2, gives:

$$t = \frac{PL}{2S - 0.2P}$$

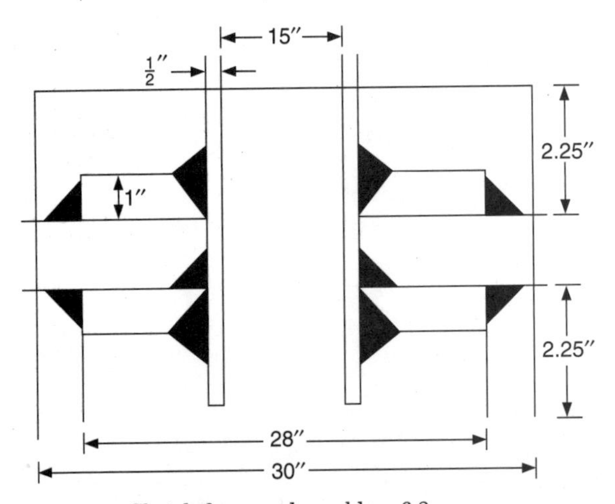

Figure 6.6 Sketch for sample problem 6.3.

where P = MAWP = 650 psig; L = inside radius = 0.9[72 – 2(1.5)] = 62.10 inches; and S = head allowable stress = 17,500 psi. Therefore,

$$t_r = \frac{650(62.10)}{2(17{,}500) - 0.2(650)}$$

$$= 1.16''$$

The required area is given in PG-33.2:

$$A = dt_rF$$

where A = required area of compensation in square inches; d = nozzle inside diameter = 15 inches (explained hereafter); t_r = required minimum thickness of head = 1.16 inches; and F = orientation factor = 1.0. Pipe above 12 NPS has its OD equal to its nominal size. Further, above 6 NPS, XS pipe has a wall of 0.5 inch. The inside diameter d of the 16 NPS XS nozzle is therefore 16 – 2(0.5) = 15 inches. We note that this nozzle could not be a manhole opening, however, because the minimum required gasket width of 11⁄16 inch (PG-44.3) cannot be achieved.

Substituting, we get:

$$A = 15(1.16)(1) = 17.36 \text{ in}^2$$

Clearly, a reinforcement pad or pads will be needed. A design as shown in sketch 6.6 is tried, with the pad thickness equal to 1 inch and all fillet welds equal to ¾ inch. The nozzle protrudes 2 inches inside the inner surface of the head. By PG-36.2, the limit of compensation measured from the center line of the nozzle along the head surface is the larger of the diameter of the finished opening (15 inches) or the radius of the opening plus the vessel wall thickness plus the nozzle wall thickness (7.5 + 1.5 + .5 = 9.5 inches). Use 15 inches. The ID of the pad is taken as 16 inches and the OD as 28 inches. By PG-36.3, the limit of compensation measured normal to the surface of the head will be the same on each side and will be the smaller of 2.5 times the head thickness or 2.5 times the nozzle thickness plus the pad thickness, or:

$$2.5(1.5) = 3.75''$$

$$2.5(.5) + 1 = 2.25''$$

The excess area within the head of the vessel is:

$$2(1.5 - 1.16)(7.5) = 5.138 \text{ in}^2$$

This area has the same strength as the missing metal, so no corrections are needed.

The required thickness of the nozzle wall (t_{rn}) is calculated from PG-27.2.2:

$$t_{rn} = \frac{PD}{2S + 2yP}$$

$$= \frac{(650)(16)}{2(15{,}000) + 2(0.4)(650)}$$

$$= 0.341''$$

The excess area in the nozzle wall outside the vessel is:

$$2(2.25)(0.500 - 0.341) = 0.717 \text{ in}^2$$

This area must be corrected downwards because of the lower allowable stress in the nozzle material. The equivalent area is:

$$0.717\left(\frac{15{,}000}{17{,}500}\right) = 0.614 \text{ in}^2$$

The area of the nozzle projecting into the vessel is:

$$2(2)(0.5)\frac{15{,}000}{17{,}500} = 1.714 \text{ in}^2$$

There are four fillet weld areas on each side of the center line, all ¾ inch size, all within the limits of compensation, and all having the allowable stress value of the pad. The area of the welds acting as compensation is:

$$4(2)(\tfrac{1}{2})(\tfrac{3}{4})(\tfrac{3}{4})\left(\frac{13{,}800}{17{,}500}\right) = 1.774 \text{ in}^2$$

The area of compensation in the pad is:

$$4(6)(1)\left(\frac{13{,}800}{17{,}500}\right) = 18.926 \text{ in}^2$$

The total area available for compensation is:

$$18.926 + 1.774 + 1.714 + 0.614 + 5.14 = 28.166 \text{ in}^2$$

There is sufficient area of compensation. We note that opening design is iterative in nature, and the current design might well benefit from further refinement. The two pads might be reduced to ¾ inch thickness, or plate of the same allowable stress value as the shell might be used, making the pad on the inside unnecessary. For the design as it is, there remains the check of the weld sizes per PW-16 and then of the paths per PW-15 and PG-37.

The joint design is most similar to Figure PW-16.1(p). The shell-to-pad fillet welds must each have a throat dimension of at least $\frac{1}{2}t_{min}$. The pad-to-nozzle fillet welds must each have a throat dimension of t_c. The minimum depth of the groove weld joining the shell to the nozzle is at least $0.7t_{min}$, where t_{min} is the smaller of ¾ inch or the thinner of two parts being joined.

$$t_{min} = \tfrac{3}{4}''$$

$$t_c = \text{smaller of } \tfrac{1}{4}'' \text{ or } 0.7t_{min}$$

$$= \tfrac{1}{4}''$$

The pad-to-shell fillet welds have a throat dimension of $0.7(\tfrac{3}{4}) > \tfrac{1}{2}t_{min}$. The pad-to-nozzle fillet welds have a throat dimension of $0.7(\tfrac{3}{4}) > \tfrac{1}{4}$ inch. The minimum depth of the groove weld joining the shell to the nozzle is 0.7(0.75) = 0.525 inch. The weld sizes therefore meet the requirements of PW-16, provided the groove weld depth is specified as (say) ⅝ inch or more.

The loads W to be carried on the load paths are calculated per PG-37.2:

$$W = dt_rS$$

$$= \{dlt_r - [(2d - dl)(t - t_r)] + A_S\}S$$

$$= \{dlt_r - [2t(t - t_r)] + A_S\}S$$

where W = the load to be carried on the load paths (use the smallest of the three values computed); dl = diameter of the unfinished opening = 16 inches; d = diameter of the finished opening = 15 inches; t_r = required thickness of the head = 1.16 inches; t = actual thickness of the head = 1.5 inches; A_S = stud hole area = zero; and S = allowable stress on the head material = 17,500 psi.

So:

$$W = (15)(1.16)(17{,}500) = 304{,}500 \text{ lb}$$

$$= \{16(1.16) - [(2(15) - 16)(1.5 - 1.16)]\}(17{,}500) = 241{,}500 \text{ lb}$$

$$= \{16(1.16) - [2(1.5)(1.5 - 1.16)]\}(17{,}500) = 306{,}950 \text{ lb}$$

Therefore, each of the load paths must have a strength at least equal to 241,500 lb. The attachment is most like Figure PW-16(c), the right side of the sketch. Load paths 1, 2, and 3 must be considered.

Load path 1 consists of the shell-to-pad fillet weld in shear and the nozzle wall in shear. No credit is taken for the groove weld connecting the pad to the shell. PW-15 gives the rules for calculating the strength of the paths. PW-15.1.2 indicates that the fillet weld should be calculated as one-half the area subjected to shear times the allowable stress times the stress factor from PW-15.2 (0.49 for fillet welds in shear).

The strength of the fillet weld is ½(π)(mean diameter of the weld)(leg size)(stress factor)(allowable stress), or:

$$\tfrac{1}{2}(3.1416)(28.75)(0.75)(0.49)(13{,}800) = 229{,}031 \text{ lb}$$

The strength of the nozzle wall in shear from PW-15.1.4 is ½(π)(mean diameter)(thickness)(stress factor)(allowable stress), or:

$$\tfrac{1}{2}(3.1416)(15.5)(.5)(0.70)(15{,}000) = 127{,}824 \text{ lb}$$

The total strength of path 1 is 229,031 + 127,824 = 356,855 lb. This is greater than the required strength of 241,500 lb, so path 1 is acceptable.

Load path 2 consists of the pad-to-nozzle fillet weld in shear and the nozzle wall in shear. The strength of the fillet weld is:

$$\tfrac{1}{2}(3.1416)(16.75)(0.75)(0.49)(13{,}800) = 133{,}435 \text{ lb}$$

The strength of the nozzle wall in shear was calculated above: 127,824 lb. The total for path 2 is 133,435 + 127,824 = 261,259 lb. This is greater than the required 241,500, so path 2 is acceptable.

Load path 3 consists of the two pad-to-nozzle fillet welds in shear plus the three groove welds in tension. PW-15.1.2 indicates that the strength of the fillet welds is 2(½)(π)(nozzle OD)(weld leg)(stress factor)(allowable stress), or:

$$2(\tfrac{1}{2})(3.1416)(16)(0.75)(0.49)(13{,}800) = 254{,}952 \text{ lb}$$

Because this is greater than the required strength, it would not be necessary to continue further with the strength calculations for path 3. The calculation is continued here to illustrate how the strength of the groove welds is calculated per PW-15.3: (2)½(π)(nozzle OD)(pad thickness)(stress factor)(allowable stress); and ½(π)(nozzle OD)(groove weld depth in shell)(stress factor)(allowable stress).

$$2(\tfrac{1}{2})(3.1416)(16)(1)(0.74)(13{,}800) = 513{,}311 \text{ lb}$$

$$\tfrac{1}{2}(3.1416)(16)(0.525)(0.74)(17{,}500) = 170{,}871 \text{ lb}$$

The strength of path 3 is the sum of the three computations shown above, 254,952 + 513,311 + 170,871 = 939,134 lb.

Openings and Compensation in Flat Heads

Openings for flat heads are discussed in PG-35. The various cases are summarized hereafter.

1. Small openings. PG-35.1 indicates that small openings are addressed in PG-32.1.3.1. This means that welded, threaded, studded, or expanded connections not exceeding NPS 2 may be installed in a flat head with no calculations required. The weld sizes must meet PW-16 for the type of welded attachment used.

2. There is no provision for using PG-32 and Figure PG-32 for flat heads. If the opening is larger than NPS 2, calculation of the compensation is required except as provided in PG-35.3, which requires that if the opening in a flat head exceeds one-half the diameter or short span of the head, then the flat head must be designed as a flange "in accordance with accepted Rules for Bolted Flange Connections." There are no such rules in Section I of the Code; Section VIII, however, has an appendix entitled "Rules for Bolted Flange Connections" that may be used (using Section I materials) in designing a flat head with an opening exceeding one-half its diameter or short span.

3. If the opening is larger than NPS 2 but less than one-half the diameter or short span, the compensation must be calculated. PG-35.2 offers three ways to provide adequate compensation:

- The required area may be computed in accordance with the formula in PG-35.2: $A = 0.5dt$, where A is the required area, d is the diameter of the finished opening, and t is the minimum required thickness of the head. (This term was called t_r with shells and formed heads.) It will be noted that the required area of compensation is only half as great for flat heads as for shells and formed heads for a given opening. This is because the basis for computing the thickness for flat heads is not the same as for shells and formed heads. The flat heads are substantially thicker to obtain the requisite stiffness and are less affected by relatively small openings. After the required area has been found, the available area in extra thickness of the head may be computed within the limits of compensation using PG-36. Other areas for compensation may include a pad and any attachment fillet welds, as well as the projection of a nozzle inside the inner surface of the head.
- The thickness of the head may be calculated by the rule in PG-35.2.1. For heads attached in a manner that does not create an edge moment, the thickness will be computed using PG-31.3, formula 1 or 3:

$$t = d\left(\frac{CP}{S}\right)^{1/2}$$

or

$$t = d\left(\frac{ZCP}{S}\right)^{1/2}$$

where d = opening diameter or short span; P = MAWP; S = head allowable stress; Z = shape factor for noncircular heads (see PG-31.3.3, formula 4); and C = attachment factor from Figure PG-31 and PG-31.4. However, the value of C is taken as twice the value from Figure PG-31 and PG-31.4, or as 0.75, whichever is less, to provide the extra thickness to compensate the opening.

- When the head is attached by bolting such that there is an edge moment, the formulae for the thickness are PG-31.3 formula 2 for a circular head and PG-31.3 formula 5 for a noncircular head. The extra thickness for the opening may be provided by doubling the value of the quantity taken to the ½ power as shown:

$$(2)(\text{modified})t = d\left[(2)\left(\frac{CP}{S} + \frac{1.9Wh_g}{Sd^3}\right)\right]^{1/2}$$

$$(5)(\text{modified})t = d\left[(2)\left(\frac{ZCP}{S} + \frac{6Wh_g}{SLd^2}\right)\right]^{1/2}$$

We note that the latter two cases state that flat heads with openings over NPS 2 but not over one-half the diameter or short span may be calculated by simply multiplying the thickness found for a blind head by the square root of 2, except that in some cases (no edge moments and $2C > 0.75$) the multiplier may be less than the square root of 2.

Sample problem 6.4. Opening in a flat head.

A flat head is 20 inches in diameter and attached to a shell as in Figure PG-31(f). An 8-inch schedule 40 pipe is inserted as a nozzle into the center of the head. The nozzle attachment is similar to that shown in Figure PW-16.1(h). All fillet welds are ¾ inch. The head thickness is 1.5 inches. For the shell, $t_s = t_r$. The allowable stresses for the shell, head, and nozzle are all 15 ksi. If the MAWP is 200 psig, what compensation is required?

Solution: The required head thickness for a blank head is given by PG-31.3, formula 1:

$$t = d\left(\frac{CP}{S}\right)^{1/2}$$

where d = head diameter = 20 inches; $C = 0.33m = 0.33(t_r/t_s) = 0.33$ per Figure PG-31(f); P = MAWP = 200 psig; S = head allowable stress = 15 ksi; and t = minimum required thickness of the head.

Therefore,

$$t = 20\left[\frac{(0.33)(200)}{15{,}000}\right]^{1/2}$$
$$= 1.327''$$

The required area of compensation is given by PG-35.2:

$$A = 0.5dt$$

where A = required area; d = diameter of finished opening; and t = minimum required thickness.

The nozzle is 8-inch schedule 40. The outside diameter is 8.625 inches. The nominal thickness is 0.322 inch, and the actual thickness may be as much as 12½ percent less. Therefore the thickness is taken as 0.282 inch, and the ID is 8.625 – 2(0.282) = 8.061 inches:

$$A = 0.5(8.061)(1.327) = 5.348 \text{ in}^2$$

The attachment welds for the head to shell joint must meet the size requirements of Figure PG-31: The weld throat must be at least 0.7 times the shell thickness. The actual throat is 0.7(0.75) = 0.525, more than enough.

The nozzle welds must meet the size requirements of Figure PW-16.1(d). Each weld throat must be at least the lesser of ¼ inch or $0.7t_{min}$. The sum of the throats must be at least 1.25 times t_{min}. From PW-16.3, t_{min} is the smaller of ¾ inch or the thinner of the parts being joined. In this case t_{min} = 0.282 inch. Each weld throat is 0.525 > ¼ inch, so the first criterion is met. The sum of the throats is 1.050 inches, which is greater than 1.25(0.282), so the second requirement is also met.

By PG-36, the limits of reinforcement are: in the direction along the head, the larger of 8.061 or 4.03 + 1.5 + 0.282; and in the direction normal to the head, the smaller of 2.5(1.5) or 2.5(0.282) = 0.705. The answers are 8.061 and 0.705, respectively.

The area available in the head is given by the first formula in PG-36.4.1:

$$A_1 = (t - t_r)d$$
$$= (1.5 - 1.327)(8.061) = 1.397 \text{ in}^2$$

The area available in the nozzle is given by the second formula in PG-36.4.2:

$$A_2 = (t_n - t_{rn})(5t_n)$$

The value t_{rn} is calculated from PG-27.2.2: t_{rn} = 200(8.625)/[(2)(15,000) + (2)(0.4)(200)] = 0.057 inch. Therefore:

$$A_2 = (0.282 - 0.057)(5)(0.282) = 0.317 \text{ in}^2$$

The area available in the welds is 4(.5)(0.75)(0.75) = 1.125 in^2.

The total compensation available is 1.397 + 0.317 + 1.125 = 2.839 in^2. This is less than the required area of 5.348 in^2. Compensation may be added, or the head thickness may be calculated per PG-35.2.1:

$$t = d\left(\frac{2CP}{S}\right)^{1/2}$$

$$= 20\left[\frac{2(0.33)(200)}{15{,}000}\right]^{1/2}$$

$$= 1.876''$$

Chapter

7

Miscellaneous Problems

There are a number of types of computations, the rules of which are given in Section I but that are not lengthy or complex enough to warrant a full chapter of this book. These situations are discussed below, with sample problems.

Safety Valves

Safety valves and their manner of operation are discussed at some length in chapter 8, Appurtenances. There are, however, some computational paragraphs in Section I that are discussed here. The relevant paragraph numbers for safety valves are: PG-67, 68, 69, 70, 71, 72, and 73; PFT-44 and Table PFT-44; PMB-15; PEB-15; PVG-12; and A-12, 13, 14, 15, 16, 17, 44 (plus Table), 45, 46, 48, 49, 310, 311, 312, 313, 314, 315, and 316.

PG-67.1 requires that each boiler have at least one safety valve and gives the conditions when two or more safety valves are needed. PG-67.2 gives the performance requirements for the safety valve(s): There must be sufficient relieving capacity to prevent the pressure in the boiler from exceeding 1.06 times the MAWP under the most severe conditions. Thus, the safety valve(s) must be capable of discharging all the steam that can be generated at the maximum firing rate without an increase in pressure to more than 6 percent above the highest valve setting, but in no case to more than 6 percent above the MAWP of the boiler. The boiler manufacturer is given the responsibility of determining the maximum steaming capacity of the boiler (PG-67.2.1), but for electric boilers there is a rule given in PEB-15 for computing the minimum relieving capacity, and for high-temperature water boilers there is a rule given in PG-67.2.4. PVG-12.6 gives a formula for calculating the required relieving capacity for an organic fluid vaporizer. Table A-44 gives steaming capacity factors for different boiler types and fuels; this table was formerly mandatory, but it has been moved to the appendix under the Code revision that simply requires the manufacturer to determine the steaming capacity of most types of boilers. This table may still be used, however, and it is particularly useful in conjunction with A-44 through A-49 for work involving boilers in service.

Sample problem 7.1. Safety valves for an electric boiler.

How many safety valves and what relieving capacity are required for an electric boiler heated by 1800 kilowatts?

Solution: PEB-15.1 requires two or more safety valves for power inputs over 1100 kW. At least two safety valves are required. PEB-15.2 requires the minimum relieving capacity to be at least 3½ lb per hour per kilowatt of heat input. The minimum capacity is therefore:

$$W = 3.5(1800) = 6300 \text{ lb/hr}$$

It is evident, then, that two valves totalling at least 6300 lb per hour relieving capacity will be required. We note that PG-71.1 permits the two valves to be mounted together on Y-bases or as duplex valves, or the two valves may be mounted separately. If mounted together, the valves must be of approximately equal capacity; if mounted apart from each other, one valve may be as much as 50 percent less in capacity than the other. Thus, in this example, one valve could have a capacity of 2100 lb per hour and the other 4200 lb per hour.

Sample problem 7.2. Safety valves for a scotch dryback boiler.

A Scotch dryback boiler has a 30-inch ID furnace 12 feet long and 52 tubes, 3.5 inches diameter by 12 feet long. The tube thickness is 0.095 inch. The boiler is stoker-fired. Find the heating surface. Find the required minimum safety valve capacity. Find the minimum number of safety valves. If the MAWP is 125 psig, find the minimum size of the opening(s) in the shell for safety valve(s).

Solution: The heating surface is computed on the side receiving heat (PG-101.1.1). The fireside area of the furnace is:

$$(\pi)(\text{ID})(\text{Length}) = 3.1416(30'')(144'') = 13{,}572 \text{ in}^2 = 94.25 \text{ ft}^2$$

The area of the tubes is:

$$52(\pi)(\text{ID})(\text{Length}) = 52(\pi)(3.5 - 2(0.095))(144) = 77{,}865 \text{ in}^2 = 540.73 \text{ ft}^2$$

The total area is:

$$94.25 + 540.73 = 635 \text{ ft}^2$$

The steaming capacity may be found by use of Table A-44, where for a stoker-fired firetube boiler the steaming rate is given as 7 lb/ft²/hr. So:

$$W = 635 \text{ ft}^2(7 \text{ lb/ft}^2\text{/hr}) = 4445 \text{ lb/hr}$$

PG-67.1 requires two or more safety valves.

PFT-44 and Table PFT-44 give the size of the openings in the shell of firetube boilers, given their MAWP and their heating surface. The formula is found in Note 1 to Table PFT-44:

$$A = \frac{HV}{420}$$

where H is the total boiler heating surface in square feet = 635 ft²; V is the specific volume of the steam at the MAWP, found in Table PFT-44 at the extreme right = 3.220 ft³/lb; and A is the required total area of the openings in the shell in square inches:

$$A = \frac{635(3.220)}{420} = 4.868 \text{ in}^2$$

From the table at the bottom of Table PFT-44, we see that we may choose standard sizes totalling the required area. We could use one 1½ NPS valve and one 2 NPS valve with areas totalling 2.036 + 3.355 = 5.391 in², or we could choose two 2-inch valves.

Sample problem 7.3. Required safety valve capacity based on firing rate.

A boiler burns pulverized coal (anthracite) at the rate of 5000 lb per hour. What is the steaming capacity (and therefore the required safety valve capacity) based on this fuel consumption rate?

Solution: A-12 gives the formula for determining steaming capacity based on fuel consumption:

$$W = \frac{0.75CH}{1100}$$

where W = steaming and relieving capacity in lb/hr; C = fuel consumption rate in lb/hr, ft³/hr, or gal/hr = 5000 lb/hr; H = heat of combustion in Btu/lb, Btu/ft³, or Btu/gal (from A-17, the heat of combustion for anthracite is 13,700 Btu/lb); the factor 0.75 is an assumed combustion efficiency (i.e., 75 percent of the heat is assumed to be used to evaporate water into steam); and the divisor 1100 is the amount of heat in Btu/lb that is assumed to be absorbed by each pound of feedwater in its conversion to a pound of steam. This would include the heat to raise the water to the boiling temperature at the operating pressure and the latent heat of vaporization.

Therefore:

$$W = \frac{0.75(5000)(13{,}700)}{1100}$$

$$= 46{,}705 \text{ lb/hr}$$

It should be noted that in the first three problems, the required safety valve capacity is a mass flow rate of steam, based on either empirical rules for steaming rates based on boiler heating surface areas or on fuel consumption rates (i.e., heat input rates), but independent of operating or relieving pressure. The steaming capacity does not change with operating pressure; what does change is the volume of the steam. Thus the size of the openings in the shell of a firetube boiler will actually be larger at lower operating pressures because the steam has a higher specific volume at lower pressures. Therefore, when a boiler is derated in service to a lower pressure, the required relieving capacity does not go down. New safety valves, with a lower set pressure, will be required, and care must be taken to see that the relieving capacity can be met through the same openings. Frequently, larger openings will be required.

Sample problem 7.4. Safety valve capacity, set pressure, and blowdown.

The boiler of sample problem 7.3 is a watertube boiler with a 650-psig maximum allowable working pressure. This boiler is fitted with a superheater and a reheat superheater. The steaming capacity was found to be 46,705 lb/hr. What safety valve capacities are required on the drum safety valves, the superheater safety valve(s), and the reheat superheater safety valve(s)? What are the highest and lowest permissible set pressures for the drum safety valves? What is the maximum blowdown for both the drum safety valves and the superheater safety valves? What is the maximum pressure in the boiler when the safety valves are blowing?

Solution: The total relieving capacity of the drum safety valves plus the superheater safety valve(s) must be capable of relieving all the steam that can be generated without the pressure going more than 6 percent above the MAWP (PG-67.2). Therefore the maximum pressure that may be experienced in this boiler with the safety valves blowing is 1.06(650) = 689 psig.

PG-68.1 indicates that the superheater must have at least one safety valve. PG-68.2 states that if the superheater has no valves that can isolate it from the boiler, its safety valves may be considered to supply up to 25 percent of the total required relieving capacity. Therefore, the superheater safety valve capacity may be 11,676 lb/hr of the total required of 46,705 lb/hr. The boiler safety valves must have at least 75 percent of the total required capacity or 35,029 lb/hr.

PG-68.4 states that the reheat superheater safety valve capacity must be at least 15 percent of the total required capacity, but this safety valve does not count towards the required aggregate total. Therefore the reheat superheater safety valve capacity for this boiler must be at least 0.15(46,705) = 7006 lb/hr.

PG-67.3 indicates that at least one drum safety valve must be set at or below the MAWP of the boiler. The highest valve setting may be 3 percent above the MAWP, and the lowest setting may be 10 percent below the highest one. Thus, the highest allowable safety valve setting on the drum will be 1.03(650) = 670 psig. The lowest setting may be 10 percent lower than the highest, or 0.9(670) = 603 psig. Thus the maximum range of set pressures for this boiler is 670 – 603 = 67 psig.

The blowdown of a safety valve is the difference between the set pressure (when it opens) and the reseating pressure (when it closes). PG-72.1 indicates that safety valves should close at a pressure not less than 96 percent of the set pressure, except that all drum valves may be set to close at 96 percent of the lowest set pressure. Therefore in this case the valves can all be set to close at 0.96(603) = 579 psig. The maximum blowdown, then, is the blowdown for the highest set valve, from 670 to 579 = 91 psi. It should be noted that the question was framed so the extreme values would be found. However, it is not necessary that the safety valves be set at these maximum and minimum settings generally.

The blowdown for the superheater and reheater safety valves may be from a minimum of 2 percent to a maximum of 4 percent of their respective set pressures (PG-72.1). The pressures for the superheater safety valves must be set such that the MAWP is not exceeded upstream from the valve. This is required in PG-68.1 and in PG-68.4 (for reheat superheaters). Therefore, the superheater safety valves will generally be set somewhat below the MAWP of the boiler.

Structural Loads on Tubes

It frequently happens that tubes are used to carry structural loads in addition to the internal pressure of normal operation. These structural loads may be the weights of boiler accessories such as soot blowers. The loads are transmitted to the tubes by lugs or brackets fillet-welded or full-penetration-welded to the outer surface of the tube. Such loads introduce stresses in the tube wall that may be additive to the longitudinal and circumferential principal stresses induced by the internal pressure. Such additional loads and stresses must be considered in designing the tubes. The rules for calculating the loads on lugs and brackets are given in PG-55 and PW-43. Sketches showing the types of loading to be analyzed are shown in Figure PW-43.2 and Figures A-71 to A-74. Sample calculations are also shown in A-71 to A-74.

Two calculations are required, independent of each other. Both are, however, based on an existing tube; the approach is therefore trial and error. The tube diameter and thickness are known, as are the allowable stress and maximum allowable working pressure. The lug is understood to be attached by welding. (PG-55 discusses the materials and manner of the welding and of calculating the strength of the fillet welds themselves. Figure PW-16.2 shows acceptable forms for attachment welds, including minimum weld sizes for the fillet welds.) The length of the bracket along the tube and the thickness are known, as are the direction and point of application of the load applied to the bracket.

One calculation is the actual load intensity, in pounds per inch, that the applied load causes in the tube at the bracket attachment. This load intensity is called the unit load in PW-43.1.1, and the formula for calculating the actual value is given in PW-43.1.2.

The second calculation is of the allowable unit load (lb/in) that the tube is capable of withstanding in service. The magnitude of the allowable unit load depends on the pressure stresses in the tube, the tube geometry, the bracket geometry, and the load direction and point of action.

Sample problem 7.5. Load on a tube bracket.

A boiler tube made of SA-192 material is 3¼ inches in diameter and 0.120 inch thick. The MAWP is 700 psig. A bracket is welded to the tube in accordance with Figure PW-16.2(a). The bracket is 5⁄16 inch thick by 4 inches long and carries a load of 1200 lb that acts away from the tube at normal incidence. The point of application is 1.5 inches from the center of the bracket. Find the minimum fillet weld sizes and check whether the structural attachment meets the requirements of PW-43.

Solution: The minimum fillet weld sizes are given in Figure PW-16.2(a): The throat must be at least $0.7t_{\min}$ but not less than 0.25 inch. PW-16.2 gives $t_{\min}$ as the smaller of ¾ inch or the thinner of the parts being joined. In this case, $t_{\min} = 0.120$ inch. The minimum fillet weld size is therefore 0.25/0.7 = 0.36 inch. A ⅜-inch fillet weld size would be specified.

The actual unit load is calculated per PW-43.1.2:

$$L = \frac{W_r}{l} \pm \frac{6We}{l^2}$$

where W_r = the load component acting perpendicular to the tube = +1200 lb (note the use of a sign convention; tension is positive and compression is considered negative); l = length of attachment to tube = 4 inches; W = load component having moment about the center of the tube attachment = 1200 lb; and e = "eccentricity of W" or the moment arm of W about the center of the tube attachment = 1.5 inches. Therefore:

$$L = \frac{+1200}{4} \pm \frac{6(1200)(1.5)}{4^2}$$

$$= +300 + 675 = +975 \text{ lb/in (tension at one end)}$$

$$= +300 - 675 = -375 \text{ lb/in (compression at other end)}$$

The second calculation step is to find the allowable unit loads in tension and compression. The actual unit loads will be compared to the allowable to determine whether the proposed design is acceptable. PW-43.2.4 gives the equation for the allowable unit load L_a in lb/in. $L_a = KL_fS_t$. K is a factor from Table PW-43.1 based on the attachment angle of the lug or bracket on the tube. L_f is the load factor, found either from Figure PW-43.1 or from equations PW-43.2.1 and PW-43.2.2. S_t is the available stress, computed from PW-43.2.3.

The attachment angle of the bracket is found either by a scale sketch or by computation: The arc length on the tube surface is about equal to the bracket thickness. Arc length = (radius)(angle in radians), so:

$$0.25 = \left(\frac{3.25}{2}\right)(\text{angle})$$

The attachment angle = 0.15 rad = 8.8 degrees.

The value of the dimensionless factor K is obtained from Table PW-43.1, based on the attachment angle. By interpolation, $K = 1.094$.

The values of the load factors are functions of the tube geometry, as expressed in the factor X.

Per PW-43.2.4, the value of X is found as $X = D/t^2 = 3.25/(0.120)^2 = 226$.

Two load factor values L_f are found, one for tension and one for compression. L_f may be found from the chart in Figure PW-43.1, using the value of X, or the values of L_f may be found by use of the equations PW-43.2.1 and PW-43.2.2. From the chart, the load factor for compression is about 0.006 and for tension about 0.007. By equation:

$$\text{Compression } L_f = 1.618X^{-1.020 - 0.014 \log X + 0.005(\log X)^2}$$

$$= 0.0063$$

$$\text{Tension } L_f = 49.937X^{-2.978 + 0.898 \log X - 0.139(\log X)^2}$$

$$= 0.0071$$

The allowable stress in the tube is found from Section II, Part D, Table 1A: S_a = 11,500 psi at 700°F. (The tube is assumed to absorb heat per PG-27.4, Note 2.) The actual stress in the tube is computed using PG-27.2.1:

$$P = S\left[\frac{2t - 0.01D - 2e}{D - (t - 0.005D - e)}\right]$$

where P = maximum allowable working pressure = 700 psig; S = allowable stress for tube material = 11,500 psi; D = tube OD = 3.25 inches; t = tube wall thickness = 0.120 inch; e = tube expansion allowance per PG-27.4, Note 4. For bracket and lug computations, e will always be zero because the added thickness, when required at all, is only needed in the part of the tube inserted into the tubesheet for expansion.

Substituting, we get:

$$S = 700\left\{\frac{3.25 - [0.120 - 0.005(3.25)]}{2(0.120) - 0.01(3.25)}\right\}$$
$$= 10{,}614 \text{ psi}$$

The available stress (PW-43.2.3) is given as:

$$S_t = 2S_a - S$$
$$= 2(11{,}500) - 10{,}614 = 12{,}386 \text{ psi}$$

The allowable unit loads can now be computed per PW-43.2.4:

$$\text{Allowable compression } L_a = KL_fS_t$$
$$= 1.094(0.0063)(12{,}386)$$
$$= 85 \text{ lb/in}$$

$$\text{Allowable tension } L_a = KL_fS_t$$
$$= 1.094(0.0071)(12{,}386)$$
$$= 96 \text{ lb/in}$$

The allowable unit loads on this tube are substantially lower than the unit load exerted by the bracket. The design does not meet the Code requirements. If the tube thickness is increased, the value of X goes down and the load factors increase. The available stress S_t also increases. Thus, if the thickness were to be increased to 0.300 inch, X would be 36, and the load factors would be about 0.04 for compression and 0.053 for tension. The available stress would increase to 19,340 psi. Then the allowable unit loads would be:

$$\text{Compression } L_a = 1.094(0.04)(19{,}340)$$
$$= 846 \text{ lb/in}$$

$$\text{Tension } L_a = 1.094(0.053)(19{,}340)$$
$$= 1121 \text{ lb/in}$$

both of which are satisfactory.

After the necessary adjustments are made, the strength of the fillet welds can be checked for compliance with PG-55. The load capacity of the welds is equal to two times the leg times the length times the allowable stress times the factor 0.55 from PG-55.2:

$$2(4)(3/8)(11{,}500)(0.55) = 18{,}975 \text{ lb}$$

This is far greater than the applied load of 1200 lb. With a 0.300-inch tube wall thickness and ⅜-inch fillet welds, the attachment meets the requirements.

Sample problem 7.6. Diagonal load on a tube bracket (sketch 7.1).

A 2½-inch tube has a 0.120-inch wall and a lug welded to it. The attachment angle is 10°. The lug is 3 inches long (along the tube) and 1 inch wide (sticking out from the tube). A load of 800 lb acts in tension on the lug at an angle of 45° below the horizontal. The load acts at the bottom outside corner of the lug. The allowable stress on the tube material is 11,500 psi. The maximum allowable working pressure on the tube is 750 psig. Does the attachment meet the requirements of PW-43?

Solution: First, the actual unit load on the tube is calculated. This is done by the principle of superposition. The 800-lb load at 45° is broken

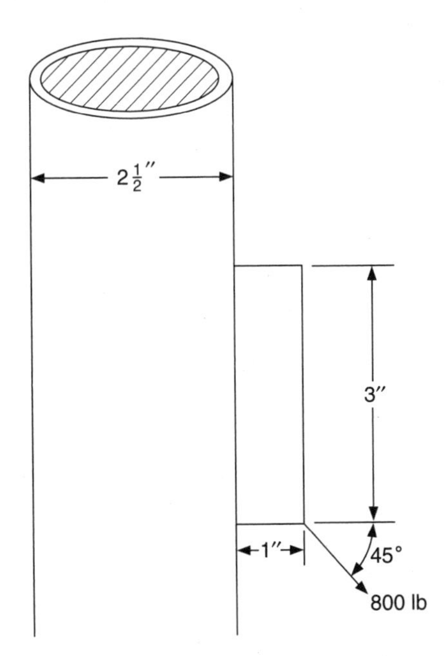

Sketch 7.1 Tube from sample problem 7.6.

into its components in the horizontal and vertical directions. The horizontal load is 566 lb acting with an eccentricity or moment arm about the center of the lug of 1.5 inches with a counterclockwise sense. The vertical component is 566 lb acting downward with a moment arm of 1 inch and a clockwise sense. PW-43.1.2 is used to compute the unit load caused by each of the components.

$$L = \frac{W_r}{l} \pm \frac{6We}{l^2}$$

hence:

$$\text{Vertical } L = 0 \pm \frac{6(566)(1)}{3^2}$$

$$= \pm 377 \text{ lb/in, tension at the top, compression at the bottom}$$

$$\text{Horizontal } L = \frac{+566}{3} \pm \frac{6(566)(1.5)}{3^2}$$

$$= +189 \pm 566 \text{ lb/in}$$

$$L = 755 \text{ lb/in tension at the bottom}$$

$$= 377 \text{ lb/in compression at the top}$$

At the top of the bracket, the sum of the tension and compression unit loads is zero. At the bottom of the bracket, the sum of the tension and compression unit loads is +755 − 377 = +378 lb/in (tension). Thus L = +378 lb/in. L_a must be calculated as before. PW-43.2.4 gives $L_a = KL_fS_t$. K is found in Table PW-43.1 for a 10° attachment angle:

$$K = 1.108$$

To find the load factor L_f we must first find X:

$$X = \frac{D}{t^2}$$

$$= \frac{2.5}{(0.120)^2}$$

$$= 173.6$$

Only the tension load factor must be found. From PW-43.2.2:

$$L_f = 49.937X^{-2.978 + 0.898 \log X - 0.139(\log X)^2}$$

$$= 49.937(173.6)^{-2.978 + 0.898 \log (173.6) - 0.139[\log (173.6)]^2}$$

$$= 0.0094$$

The available stress S_t is found from PW-43.2.3:

$$S_t = 2S_a - S$$

where S_a was given as 11,500 psi.

S is computed by use of PG-27.2.1 as 8,346 psi, so:

$$S_t = 2(11{,}500) - 8{,}346 = 14{,}654 \text{ psi}$$

Then, by PW-43.2.4:

$$L_a = 1.108(0.0094)(14{,}654) = 153 \text{ lb/in}$$

The lug attachment does not meet the requirements of PW-43.

Proof Testing

The ASME Boiler Code provides rules for computing common shapes under common pressure loadings, but there may arise situations where a component cannot be calculated by the rules of Section I. Other calculations may be used if both the Manufacturer and the Authorized Inspector (AI) are satisfied that such calculations adequately describe the stresses in the component with "a reasonable degree of accuracy" (PG-18). If there are no rules in the Code and if it is "impossible to calculate with a reasonable degree of accuracy . . . ," then a proof test is required. It should be noted that PG-18 does not preclude the use of computational methods not in the Code, such as finite element analysis, provided that such methods have demonstrated validity for the particular application. PG-100 repeats PG-18 verbatim. The methods of performing proof tests are given in A-22. It is important to note that proof testing can only be used for parts that cannot be calculated. Further, proof testing cannot be used to achieve a higher MAWP than the calculated MAWP on a part that can be calculated (A-22.1). In addition to the general conditions described in PG-18, proof testing is especially mentioned in PG-32.3.3 as being advisable in the case of "very large openings . . . or openings of unusual shape."

There are several types of proof tests, classified by test method, materials restrictions, and external vs. internal pressure loading on the part. Parts subject to external pressure ("subject to collapse") are proof-tested in accordance with A-22.7. The part must withstand a hydrostatic test pressure of three times the maximum allowable working pressure "without excessive deformation." Some localized yielding may be permitted, but no significant dimensional changes will be acceptable. A-22.10 requires that proof tests to establish MAWP be witnessed and approved by the Authorized Inspector. Therefore, "excessive deformation" as a practical matter is in the judgment of the AI. The Manufacturer, however, remains responsible for parts designed by the proof test when these parts enter service; this party is therefore well advised to adopt a conservative definition of "excessive deformation" for parts under collapse. The test pressure must be corrected for temperature if the allowable stress in the material is different at the service temperature than the test temperature (A-22.7.2). The methodology for correcting for temperature differences and corresponding allowable stress differences is given in A-28.

Sample problem 7.7. Proof test pressure for a part subjected to external pressure.

What is the proof test pressure for a part under collapse if it is made of SA-285, grade C material and is intended for a maximum allowable working pressure of 250 psig at 800°F?

Solution: A-22.8 gives the formula for correcting the MAWP at the test temperature to the MAWP at the operating temperature:

$$P_o = \frac{P_t S_o}{S_t}$$

where P_o = the MAWP at the operating temperature = 250 psig; P_t = the MAWP at the test temperature; S_o = the allowable stress at temperature = 10,200 psi (Section II, Part D, Table 1A); and S_t = the allowable stress at the test temperature (assumed ambient) = 13,800 psi. So:

$$250 = P_t\left(\frac{10{,}200}{13{,}800}\right)$$

$$P_t = 338 \text{ psig}$$

Then, for a 338-psig MAWP at ambient temperature, the test temperature must be three times higher (A-22.7.1):

$$\text{Test } P = 3(338) = 1015 \text{ psig}$$

We note that this is simply three times the MAWP at temperature times the stress ratio S_t/S_o. Given an ambient test pressure, one would divide by 3 and multiply by S_o/S_t.

Components subject to internal pressure may be proof-tested in a number of ways, depending on the materials of construction and the wishes of the Manufacturer. Burst testing may be used for any component subjected to internal pressure that requires proof testing and is made of any Section I material. This method is therefore the most broadly applicable of the proof tests for internal pressure. The part is subjected to a hydrostatic test with pressure being increased until the part fails. The burst-test pressure is recorded and used to compute the MAWP of the component. The formula used depends on the material of construction as shown in A-22.6.3.2.1, 2, 3, and 4.

Sample problem 7.8. Burst tests to determine MAWP.

A hydrostatic burst test is conducted on a component and the part burst at a test pressure of 750 psig. Find the maximum allowable working pressure at 300°F if the test was conducted at 75°F and the part was made of:

(a) SA-285 grade B steel. Four samples of the plate were tensile-tested and the average ultimate strength was 58,500 psi.
(b) SA-278 class 45 gray cast iron. The test bar tensile test resulted in a UTS of 46,500 psi.
(c) SA-395 nodular iron. The test bar broke at 64,000 psi.
(d) SA-216 cast steel grade WCA.

Solution: The burst test pressure is used with the formulae in A-22.6.3.2:

(a) The MAWP is the bursting pressure divided by 5. However, a correction factor is applied if the actual strength of the material is greater than the specified minimum strength. (If the design is approved, a later component may be built from material which just meets the minimum tensile strength specified.)

$$P = \left(\frac{B}{5}\right)\left(\frac{S}{S_a}\right)$$

where P = MAWP; B = burst pressure from test = 750 psig; S = specified minimum tensile strength for the material from Section II = 50,000 psi; and S_a = Actual average tensile strength from tests of material = 58,500. Then:

$$P = \left(\frac{750}{5}\right)\left(\frac{50{,}000}{58{,}500}\right) = 128 \text{ psig}$$

(b) For gray cast iron, A-22.6.3.2.2 gives:

$$P = \left(\frac{B}{6.67}\right)\left(\frac{S}{S_b}\right)$$

where P = MAWP; B = bursting pressure; S = specified minimum tensile strength = 45,000 for class 45 iron; and S_b = actual strength of test bar. It should be noted that the Code contains an error in A-22.6.3.2.2 by defining S_b as the "minimum tensile strength of the test bar." The minimum strength of the test bar is, however, the minimum strength of the cast iron. The intent of the Code is again to correct the burst test results from a strong batch of material to allow for the fact that weaker materials meeting the specification may be used in production parts:

$$P = \left(\frac{750}{6.67}\right)\left(\frac{45{,}000}{46{,}500}\right) = 109 \text{ psig}$$

(c) For the nodular iron component, A-22.6.3.2.3 gives:

$$P = \left(\frac{Bf}{5}\right)\left(\frac{S}{S_b}\right)$$

All the terms except f have been previously defined; f is a casting quality factor "defined in PG-25." Thus, although PG-25 applies to steel castings, it is incorporated into the proof test burst-test requirement for nodular iron. The quality factor will be 0.8. So:

$$P = \left(\frac{750}{5}\right)(0.8)\left(\frac{60{,}000}{64{,}000}\right) = 113 \text{ psig}$$

(d) For the cast steel, A-22.6.3.2.4 gives:

$$P = \left(\frac{Bf}{5}\right)\left(\frac{S}{S_a \text{ or } S_m}\right)$$

In this case we do not have the actual tensile test results for the material. The specification gives a tensile test UTS range of 60,000 to 85,000 psi for the SA-216 grade WCA material. Therefore, we use $S_m = 85{,}000$ in the formula. Again a quality factor of 0.8 is applied (since we have no basis for using a higher value—see PG-25). Substituting, we get:

$$P = (750)\left(\frac{0.8}{5}\right)\left(\frac{60{,}000}{85{,}000}\right) = 85 \text{ psig}$$

It should be noted that if any of the materials in parts (a) through (d) have lower allowable stresses at the operating temperature than at the test temperature (again assumed ambient), then the MAWP calculated must be adjusted (i.e., reduced) by multiplying by the stress ratio as per A-22.8.

For some materials, the MAWP of a component may be determined or obtained by a hydrostatic test to the yield point of the material. The material must have a yield point or yield stress that is sufficiently far from the ultimate strength of the material. A-22.2.1.1 requires that the specified minimum yield be no greater than five-eighths of the specified minimum ultimate strength if proof testing based on yielding is to be permitted. (Otherwise burst testing must be done.) The measurement of yielding may be accomplished by use of strain gages applied to the surface of the component at critical locations, or by use of dial indicators or other displacement measuring devices to measure overall changes in the dimensions of the part. In either case, care must be used to be certain that the strain or displacement measuring devices are so placed as to record the maximum values. Brittle coating testing may be used to identify the areas where strain gages should be applied.

In both types of testing, the pressure is applied in increments (see A-22.3.2). At each pressure step, the readings of all the strain gages, dial gages, or other devices are recorded. The pressure is then brought to zero and the gage/dial readings are again recorded. It is useful to plot the curves of strain or displacement vs. pressure and values of the gages at zero pressure as the test progresses. The "answer" in these tests is the value of pressure at which the strain-pressure curve or displacement-pressure curve has become nonlinear. This should also be the point at which the gages begin to fail to return to zero when pressure is removed. The distance from zero should increase with each reading once the yield pressure is reached. Once a value of the yield pressure H is obtained, the MAWP for the component can be computed by A-22.6.1.4.1 or A-22.6.1.4.2 for the strain gage test.

The first formula is used when tensile test data has been obtained by testing samples cut from the tested component in areas where the yield point was not reached or exceeded. The value of the MAWP from the test is one-half the yield pressure, adjusted by multiplying by the ratio of the minimum yield strength divided by the actual yield strength. When tensile test results are not available, the MAWP from the strain gage proof test is simply 40 percent of the yield pressure (A-22.6.1.4.2).

For the displacement test, A-22.6.2.4 gives three formulae for computing the MAWP, once the value H has been obtained from the test data. When tensile test data has been obtained, the formula is the same as in the case of the strain gage test. When no tensile test data is available for the component material, the MAWP may be either 40 percent of H or, if the material is carbon steel and has a specified minimum UTS of not more than 70,000 psi, $H/2$ times the ratio of the specified minimum UTS to the UTS + 5000.

Sample problem 7.9. Proof test by yielding.

A proof test is conducted and the pressure vs. strain data shows the onset of yielding took place at 860 psig. Find the MAWP if:

(a) The test was a strain gage test. The material had a minimum specified yield strength of 40,000 psi and a minimum specified ultimate tensile strength of 65,000 psi.
(b) The test was a tensile test, material specification as in (a). The average yield strength of the material was determined to be 41,700 psi.
(c) The test was a displacement test. Same material specification as in (a).

Solution: In (a), the formula is given in A-22.6.1.4.2:

$$P = 0.4H$$

$$= 0.4(860) = 344 \text{ psig}$$

In (b), the formula is given in A-22.6.1.4.1:

$$P = \frac{0.5HY_S}{Y_a}$$

$$= \frac{0.5(860)(40{,}000)}{41{,}700} = 412 \text{ psig}$$

In (c), if we had the same tensile test data, we would have the same answer as in (b). With only the material data, we may choose between the result in (a) and that given by formula A-22.6.2.4.2.1:

$$P = \frac{0.5HS}{S + 5000}$$

$$= \frac{0.5(860)(65{,}000)}{65{,}000 + 5000}$$

$$= 399 \text{ psig}$$

The MAWP for part (c) is therefore 399 psig.

Chapter

8

Appurtenances

Section I of the Code contains requirements that each boiler be equipped with certain appliances to ensure safe operation. These requirements and the various appliances or appurtenances are discussed in this chapter.

Safety Valves

Starting at the top of the boiler, the first and perhaps the most important safety appliance on the boiler is the safety valve. Paragraphs PG-67 to PG-73 contain the requirements for safety valves. In the early days it became recognized that a device was needed to ensure that the operating pressure did not exceed the safe working pressure of the boiler. The first safety valves were of the weight-and-lever design (photo 8.1). This type of safety valve had many failings. It was often found that the lever provided a way to *increase* the pressure by applying new weight. Many times the valve was simply a convenient place for the engineer to hang his lunch pail or coat. This often resulted in disaster. The weight-and-lever safety valve or dead weight safety valve was outlawed by some jurisdictions as early as 1921 (e.g., the Massachusetts Steam Boiler Rules, Form U). These devices are now specifically prohibited by PG-67.5. Nonetheless, weight-and-lever valves were still in use in marine service into the 1940s.

In safety valve design evolution, the weighted lever on these valves was replaced by a spring. The disc was improved, and the valve incorporated an increased area as soon as it started to lift (the "huddling chamber"). This resulted in a valve that sharply sprang to full opening—the pop safety valve.

Today we have a very well-defined set of rules for the design criteria, application, and approval of safety valves and safety relief valves to pro-

Photo 8.1 A weight and lever safety valve. Note the holes in the lever for "fine adjustment." Photo by the authors.

tect boilers from overpressure. We are able to select a valve or valves to meet our needs from any number of manufacturers that have designed and tested their safety valves and safety relief valves in accordance with the requirements of Section I of the Code. Each manufacturer uses certain features of its design that differ from the others to meet the Code requirements. Valve makers such as Crosby Ashton, Crane, Lunkenheimer, and Consolidated manufacture a range of valves to meet the pressure and temperature requirements of the different types of boilers.

Safety valves, safety relief valves, and relief valves that have been designed and rated in accordance with these requirements and that have been flow-tested in accordance with paragraph PG-69 are then certified to bear the ASME code symbol. As indicated earlier, safety valves are generally referred to as the spring-loaded pop type. This type of valve is used for the relief of steam (or other vapor). Safety relief valves are designed and rated to relieve either steam or water, while relief valves are designed specifically for liquid service.

A safety valve, safety relief valve, or relief valve functions automatically when the pressure in the boiler or superheater exceeds the set pressure of the valve. When the set pressure is reached, the safety valve opens almost to full lift with a pop, and then, as the pressure increases, the valve opens to the full lift and maximum flow within 3 percent above the set pressure of the valve (PG-72.1). As the pressure is reduced below the set pressure (by an amount called the blowdown), the safety valve closes rapidly, thus preventing damage to the seat and disc by steam cutting or erosion. Relief valves, by contrast, open gradually as the pressure exceeds the set pressure. Safety valves for steam service may have a bonnet that encloses the spring, or the spring may be open, exposed to the ambient temperature. The open-spring-type may be used for either saturated or superheated steam (photo 8.2), but it is required to be used on superheated steam service (PG-68.6).

There are certain fundamentals that all safety valve manufacturers must include in their designs:

1. The valve must be provided with a spring made of a material suitable for the temperature.
2. There must be a method of adjusting the spring compression pressure to resist the desired relieving pressure in the boiler or other components of the steam system. The opening pressure shall not be adjusted more than 5 percent above or below the pressure stamped on the valve (PG-72.3).
3. There must be a method of manually lifting the disk from the seat to verify that it is free to operate. This lifting lever should be sufficiently strong to compress the spring when the pressure is at least 75 percent of the set pressure of the safety valve (PG-73.1.3).

Photo 8.2 A pop safety valve with exposed spring, suitable for saturated or superheated steam service. Note the elbow, discharge pipe, and drain. Photo by the authors.

4. There must be a method of adjusting the blowdown pressure of the valve. Proper adjustment of the blowdown assures full valve lift and consistent opening and closing. If the blowdown is not sufficient, the valve may chatter and steam draw, causing damage to the seat and disc. The blowdown shall not exceed 4 psi up to 100 psi and 4 percent of the set pressure above that (PG-72.1). On forced-flow steam generators and high-temperature hot water boilers, the blowdown shall not exceed 10 percent of the set pressure. Such valves must be marked so that they will not inadvertently find their way into normal boiler service.
5. All components such as the seat and disc must be securely fastened so that they cannot come loose and obstruct the discharge of the valve (PG-67.5).
6. All safety and safety relief valves certified for use on steam boilers, superheaters, economizers, and hot water heating boilers must be provided with a name plate (PG-110 and PG-73.3.5) containing:
   (a) the name of the manufacturer or assembler.
   (b) the manufacturer's design or type designation or number.
   (c) the size of the inlet NPS (nominal pipe size).
   (d) set pressure, psi, and capacity in lb/hr.
   (e) the date or date code of the manufacturer or assembler.
   (f) the ASME symbol.

The name plates of most safety valves also contain the seat diameter and lift.

Each boiler must be provided with a safety or safety relief valve or valves having sufficient capacity to relieve all the steam the boiler can generate (PG-70). The safety valve or valves on the superheater may be included when determining the relieving capacity of the boiler. The superheater safety valve or valves may only account for a maximum of 25 percent of the total required relieving capacity of the boiler, even though the actual superheater safety valve capacity may be greater than 25 percent of the required boiler relieving capacity (PG-68.2). It is also required that there be no valves between the boiler and superheater for PG-68.2 to apply.

Safety valves on superheaters having steam temperatures over 450° shall be made of steel, a steel alloy, or other heat-resistant material. As noted above, the spring shall be exposed to the air to protect it from the heat. See PG-68.6. The purpose of the safety valve or valves on the superheater is to provide a positive flow of steam through the superheater elements at all times. This protects the superheater from being damaged by overheating, but this protection is only assured if the safety valve or valves on the superheater are set to blow below those on the boiler proper, so a positive flow of steam is maintained when the safety valves are blowing. It is necessary when selecting and setting the superheater safety valves that consideration be given to the loss of pressure as the steam passes through the superheater (PG-68.1). This pressure loss is considerable and will in most cases exceed the allowable adjustment of the spring on the safety valve if it is purchased with the same set pressure as the drum valves.

Reheaters shall have one or more safety valves with an aggregate relieving capacity equal to the steam flow through the reheater (PG-68.4). As with superheater safety valves, one safety valve on the reheater shall be located between the outlet of the reheater and the first stop valve. This valve should have a relieving capacity of at least 15 percent of the relieving capacity required for the boiler and be set to blow at a pressure below that of the other valves on the reheater to ensure flow.

The smaller boilers (those with less than 500 square feet of heating surface) or electric boilers with a power input of less than 1100 kW require but one safety valve set at or below the maximum allowable working pressure (PG-67.1). All boilers having heating surfaces or input power more than that stated above require two or more safety or safety relief valves.

Safety valves should be selected for the particular application and operating conditions. In the event the boiler only requires one safety valve, the valve should have sufficient capacity to discharge all the steam that the boiler manufacturer has rated the boiler to produce. In the event that it is necessary to determine the generating capacity of the boiler in pounds of steam per hour, the guidelines of paragraph A-44 should be followed. To do so, it is necessary to calculate the number of square feet of boiler heating surface, including any extended heating surfaces. This value should then be multiplied by the values given in Table A-44 for the fuel being fired and the method of firing. The values determined in this method may not be adequate in all cases, and a greater safety or safety relief valve discharge capacity may be necessary. Regardless of the calculated results, the aggregate relieving capacity of the safety or safety relief valve or valves shall not be less than the manufacturer's designed maximum steaming capacity.

When the generating capacity of a boiler cannot be reliably calculated or otherwise determined, an accumulation test should be conducted in accordance with the requirements of A-46.1. The test consists of firing the boiler at the maximum rate with all steam outlets closed and observing whether the safety valves keep the pressure from exceeding the MAWP by more than 6 percent. An accumulation test should only be conducted by experienced boiler operators or engineers, exercising extreme care to be sure the boiler is not overpressurized. Since superheaters and reheaters require a flow of steam to prevent damage by overheating, boilers equipped with these devices should not be subjected to an accumulation test. It is also not recommended that high-temperature water boilers be subjected to an accumulation test. This is because of the danger of testing with hot water. However the generating capacity of the boiler has been established, the safety or safety relief valve or valves shall have a minimum relieving capacity at least equal to the generating capacity of the boiler. The capacity stamped on a valve having the ASME symbol is the manufacturer's guarantee that the valve complies with the ASME requirements for the details of construction. See PG-69.5. Whenever the method of firing or the fuel being fired is changed or additions to the heating surface are made, the safety valve relieving capacity should be rechecked.

It should be emphasized that although the safety valve is primarily used to protect the boiler from overpressure, it may also be used to keep the pressure below the maximum allowable to protect other equipment that is being supplied from the boiler when other means such as reducing valves and low-side safety valves are not provided.

When two or more safety or safety relief valves are required on a boiler, it is not desirable to have all of the valves set at the same pressure. It is also not desirable to have too broad a range of settings, since damage may occur to the valves set to the lower pressures. With the exception of a forced-flow steam generator with no fixed waterline that is protected in accordance with PG-67.4 and 67.4.1, the safety valve shall be set in accordance with PG-67.3. At least one safety valve on the boiler shall be set at or below the maximum allowable working pressure of the boiler. The additional valves may be set above or below the MAWP, but in no case shall a valve be set more than 3 percent above the MAWP. The total range of settings of all the valves shall not exceed 10 percent of the pressure of the highest set valve. With all safety valves blowing, the boiler pressure should not exceed 6 percent above the setting of the pressure of the highest set safety valve, but in no case should the pressure exceed 6 percent above the MAWP.

Safety valves should be designed, constructed, and adjusted so that the opening is sharp and the valves do not simmer or chatter. The valves should reach their maximum lift or opening at a pressure not to exceed 3 percent above the set pressure. As the pressure is reduced, all valves must close at a pressure not lower than 96 percent of their set pressure. In other words, each valve should be closed when the boiler pressure has been reduced to not less than 4 percent below the set pressure of that valve (PG-72.1). The exception to this rule is that all of the valves installed on the drum of a single boiler may be set to reseat at a pressure not less than 96 percent of the setting of the lowest set valve (PG-

67.3 and PG-72.1). This latter condition can only be accomplished by increasing the blowdown of the valves set at the higher pressures.

There are somewhat different rules applying to forced-flow steam generators with no fixed waterline (in other words, boilers operating above the critical temperature and pressure). The safety valves on such boilers may be set to reseat at a blowdown of not more than 10 percent of the set pressure. This rule also applies to high-temperature water boilers.

The Code also recognizes that it is not practical to set the popping or relieving pressure to the exact pressure desired or stamped on the valve. The acceptable range of popping or relieving pressure is referred to as the popping point tolerance and is given as stated below and in PG-72.2.

Set pressure	Tolerance
70 and below	2 psi
Above 70 to and including 300	3% of set pressure
Above 300 to and including 1000	10 psi
Above 1000	1% of set pressure

When designing the boiler, the method of mounting the safety valve or valves should be given careful consideration. Each valve may be mounted directly on the drum, or Y-type connections or duplex valves may be used. When two valves are singly mounted and have different relieving capacities, the smaller valve must have a capacity of at least 50 percent of the valve with the greater capacity (PG-71.1). When not mounted directly on the boiler, the valves should be connected to piping as close to the boiler as practical. Safety valves should be mounted in a vertical position, and there must never be any valves between the boiler and the safety valves. See PG-71.1, 71.2, and sketch 8.1.

The inlet opening to the safety valve should never be smaller than the nominal size of the valve. In fact, the inlet should preferably be about ½ inch larger to prevent pressure drop to the valve. When discharge piping is used to lead the steam out of the boiler room, the piping should not be directly connected to the safety valve. It is recommended that a long-radius elbow be attached to the discharge of the safety valve with a drip pan near the end of the elbow. The vertical discharge pipe is placed down over the end of the elbow and extends below the top of the drip pan. Care should be taken to leave sufficient space below the end of the discharge pipe to prevent it from bottoming against the drip pan when it expands. Due to the volume change of the steam passing through the valve, the discharge pipe must be carefully sized so a back pressure is not created in the safety valve discharge.

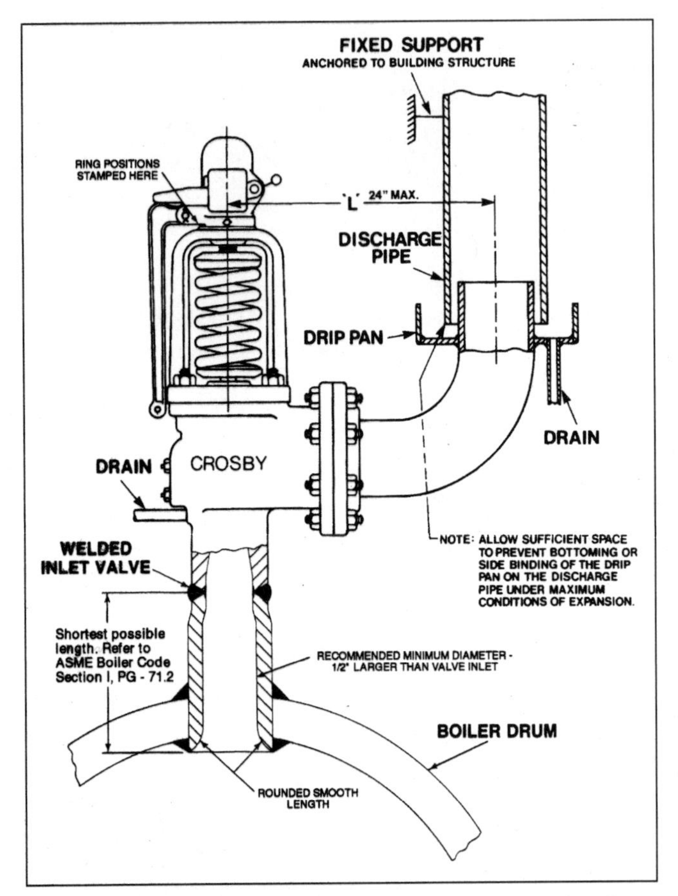

Sketch 8.1 Recommended safety valve installation practice. Courtesy Crosby Valve & Gage Company.

Such a back pressure may cause the valve to cycle and chatter. It should also be taken into account when sizing the discharge pipe that the relieving capacity stamped on the valve is only 90 percent of the actual relieving capacity. These factors should also be considered when designing a muffler to be used on the discharge piping. The muffler must also be designed so that there is no back pressure generated when the safety valve discharges.

It should be recognized that the ASME Code does not carry the force of law in all states, provinces, or countries relying on it for guidance, unless the Code and all its requirements have been adopted by those Jurisdictions. Some Jurisdictions adopt parts of the Code and not other parts, and the Jurisdiction is the ruling authority over boilers once they are fabricated and installed. With this in mind, the reader's attention is directed to PG-72.3, which states: "If the set pressure is to be adjusted within the limits above, the adjustment shall be performed by the manufacturer, his authorized representative, or an assembler." This rule has not been universally adopted. Some Jurisdictions allow the operating engineer to adjust the safety valves for set pressure and blowdown and further require the setting to be witnessed by the Inspector. An example of this is found in the *Synopsis,* which gives the rule for the state of Maine as follows: "Adjustment to safety and safety relief valve set pressure and blowdown shall be performed under the supervision of an organization in possession of a certificate of authorization or may be adjusted on the boiler in the presence of an Authorized Inspector and sealed." The National Board Inspection Code also permits setting of the popping point pressure and blowdown of safety valves by persons authorized to do so by the Jurisdiction.

When it becomes necessary to adjust a safety valve in the field, one must be sure that the rules and regulations applicable to the area or Jurisdiction are followed. In any event, power plant operators and engineers as well as Inspectors should understand the construction, operation, and adjustment of safety and relief valves.

Before a safety or safety relief valve is installed on a boiler, it should be carefully examined. The name plate should be verified as to relieving capacity and set pressure. All tags attached to the valve should be read very carefully. Some safety valves are shipped with a test plug installed in the nozzle. This plug is used in conjunction with a gag when applying the hydrostatic test to a pressure of 1.5 times the MAWP of the boiler. The plug and gag must be removed before the boiler is placed in service. The person or Inspector responsible for conducting or witnessing the hydrostatic test should also be responsible for removal of the gags and any test plugs used before allowing the boiler to be put in service. The importance of this precaution cannot be overemphasized; the authors, when examining one of four boilers in a plant, found gags on all valves on all the boilers, including those being steamed. The gags had been in place since the boilers were originally installed, about 20 years before our inspection.

One of the main troubles that develops with a safety valve during use is leakage, which occurs after the valve has blown or been tested. This is often caused by particles of rust or scale lodged between the seat and disc. Usually this can be corrected by lowering the pressure below the popping point and lifting the disc of the seat with the lifting lever and giving the valve a good blow. This will generally clean the seat areas, and the valve should reseat properly. Leakage caused by erosion or steam cutting of the seat or disc can only be corrected by removing the valve from the boiler and having it repaired by the manufacturer or assembler. Leakage may also be caused by strains applied to the valve by incorrectly installed or supported piping. This is often caused by the discharge pipe bottoming out in the pan in the discharge line. The correction is to properly install and support the piping.

If, during operation, the popping pressure changes, the valve should be removed and repaired by the manufacturer, or repair organization. If, however, it is deemed necessary to adjust the popping pressure, either to limit the pressure in the boiler or to sequence the valves, this may be done without taking the valves off the boiler. The sequenced adjustment may not exceed either 5 percent above or below the pressure stamped on the valve. If the valve chatters or cycles too frequently, the adjustment of the blowdown should be checked and corrected as needed.

Safety valves may have either two or three adjustments. All valves have at least two adjustments: The first is the popping pressure adjustment, and the second is the blowdown adjustment. The third adjustment is the nozzle ring, which is set and sealed at the factory and should not be changed (see sketch 8.2).

In adjusting the set pressure of a safety valve, the disc should not be rotated on the seat; therefore the adjustment should be made while the valve is blowing. To make the adjustment, install a recently calibrated pressure gage on the boiler for the purpose of resetting the safety valve to a correctly known pressure. Remove the seal and pin in the fork of the lifting lever. Remove the lifting lever and cap. Loosen the lock nut, raise the boiler pressure until the valve pops, and then make the adjustment. Screwing down on the adjusting screw increases the pressure, and vice versa. After making the adjustment, bring the pressure up and check the popping pressure. Be sure that the change in the setting does not exceed 5 percent of the pressure stamped on the valve. Retighten the lock nut, and replace the cap and fork. Replace the pin and reseal it.

A change in the relieving pressure of the safety valve changes the blowdown, so this may also have to be reset in order for the valve to operate properly.

To adjust the blowdown, it is necessary to break the seal on the adjusting-ring set screw, often referred to as the plug. This is the upper screw. The screw should be removed to expose the notches in the ring. With a screwdriver blade inserted into the opening, rotate the ring as required. Moving the notches to the right raises the ring and reduces the blowdown. Rotating the ring to the left lowers the ring and increases the blowdown. Care should be taken not to raise the ring too much, because if the ring is too high, it will not control the valve. This will reduce the initial lift and cause the closing to be indistinct. If the safety valve is operated this way, the seat and disc may be damaged. To correct this condition, move the ring down until the opening of the valve is crisp. The ring should not be moved more than eight to ten notches without retesting. When the blowdown is properly adjusted,

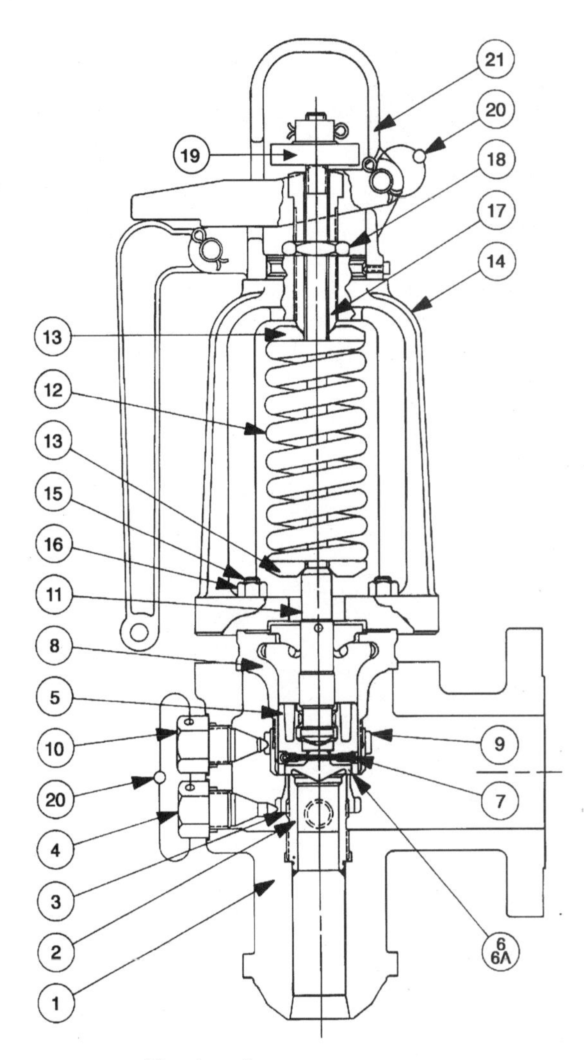

Sketch 8.2 *(Continued)*

replace the screw, being careful to see that the point goes into the notch and does not set on the top of a tooth. The adjustment screw should then be resealed.

When the above adjustments are being done on a boiler having two or more valves, the valves not being tested should be gagged to prevent them from blowing repeatedly during the setting of the other valves. When a gag is applied, the screw should be finger-tight. Under no circumstances should a wrench be used, and the person setting the safety valve should remove the gags.

Part Ref. No.	Material and Maximum Temperature			Spare Parts Designation (See Notes 1,2,3)
	Part Name	750F (399C)	1050F (566C)	
1	Body	Carbon Steel SA-216 Gr WCC	Alloy Steel SA-217 Gr WC6	
2	Nozzle	Stainless Steel	Stainless Steel	3**
3	Nozzle Ring	Stainless Steel	Stainless Steel	3
4	Nozzle Ring Set Screw	Stainless Steel	Stainless Steel	
*5	Disc Holder	Copper Nickel Alloy	Copper Nickel Alloy	2
*6	Disc Insert	Stainless Steel	Stainless Steel	1
*6A	Disc	Stainless Steel	Stainless Steel	1
*7	Disc Insert Cotter	Stainless Steel	Stainless Steel	1
8	Guide	Copper Nickel Alloy	Copper Nickel Alloy	3
9	Guide Ring	Stainless Steel	Stainless Steel	3
10	Guide Ring Set Screw	Stainless Steel	Stainless Steel	
11	Spindle Assembly	Stainless Steel	Stainless Steel	3
12	Spring	Alloy Steel Corrosion Resistant Coating	Alloy Steel Corrosion Resistant Coating	
13	Spring Washers	Steel	Steel	
14	Bonnet	Carbon Steel SA-216 Gr WCC	Alloy Steel SA-217 Gr WC6	
15	Bonnet Studs	Alloy Steel SA-193 Gr B7	Alloy Steel SA-193 Gr B16	
16	Bonnet Stud Nuts	Steel SA-194 Gr 2H	Steel SA-194 Gr 2H	
17	Adjusting Bolt	Stainless Steel	Stainless Steel	
18	Adjusting Bolt Nut	Stainless Steel	Stainless Steel	
19	Spindle Nut	Stainless Steel	Stainless Steel	
20	Seal & Wire	Lead & St. St.	Lead & St. St.	
21	Cap/Lever Assembly	Steel/Iron	Steel/Iron	

*One piece disc (6A) replaces Part Reference Numbers 5, 6 and 7 in 1-1/2" and 2" inlet sizes.
**Not a spare part for welded inlet valves.

Sketch 8.2 *(Continued)*

When all adjustments have been made and completed, the valves should be pressure-tested to verify the settings and then resealed and stamped with the new setting.

Although supercritical forced-flow steam generators are not in demand, they may still be manufactured. The rules for safety valves for such boilers are given in PG-67.4. These boilers may be equipped with power-activated safety valves with a capacity of not less than 10 percent of the maximum generating capacity of the boiler (PG-67.4.1). An outside screw and yoke-type valve may be installed between the boiler and the power-actuated relief valve to permit isolation and repair of the relief valve. The boiler shall also have enough safety valve capacity, including the power activated valves, to relieve all the steam the boiler can generate without exceeding the maximum allowable working pressure by more than 20 percent. The power-activated valves cannot account for more than 30 percent of the total safety valve capacity.

There are also special rules given in PVG-12 for safety valves on organic fluid vaporizers. Such safety valves shall not have a lifting lever and shall discharge through piping to a safe place outside the building or to a vapor condenser. Drains are not required on safety or safety relief valves used on organic fluid vaporizers. Rupture discs may be used between the vaporizer and the safety valve to protect the safety valve from corrosion and prevent leakage. When a rupture disc is used, a valved drain connection, pressure gage, or vent should be provided between the rupture disc and the safety valve so it can be used to determine whether the disc has failed. Because these safety or safety relief valves are not tested periodically, they must be removed from the vaporizer annually for disassembly, cleaning, reassembly, and testing. The rupture disc should fail below the set pressure of the safety valve and should not restrict the relieving capacity of the safety valve.

Safety valve requirements for miniature boilers are contained in PMB-15 (minimum size NPS ½). Electric boilers are treated in PEB-15.

Pressure Gage

For safe operation, each boiler must be provided with a pressure gage with a range of not less than 1½ times the maximum allowable working pressure of the boiler. The desired pressure range is approximately twice the MAWP (PG-60.6.1). A steam gage shall be connected to the steam space, or the water column, or it may be attached to the steam connection to the water column. Steam gages shall be provided with a siphon or other arrangement so that steam will condense, thus forming a water seal and preventing steam from entering the gage (see PG-60.6.1 through PG-60.6.4). There shall be a cock or valve in the gage line so that it can be shut off to replace the gage. A connection shall also be provided so that a test gage can be installed to permit checking the boiler gage (PG-60.6.3).

The gage connections to the boiler, if used, may be of brass or copper as long as the temperature does not exceed 406°F. The piping must be at least ¼ inch standard pipe size. The siphon, if used, should also have an inside diameter of at least ¼ inch.

Where the piping to the gage is of steel or wrought iron, the size of the piping shall be at least ½ inch inside diameter to prevent the chances of becoming plugged by corrosion. It is permissible to install a second valve or cock in the gage line, close to the boiler, provided that it is sealed or locked in the open position. The piping to the pressure gage must be sized and laid out so that it can be blown out to clear the line. A valve should be located at the end of the line to facilitate this process. Refer to PG-60.6.1.

On forced-flow steam generators with no fixed water line, the generator shall be provided with a pressure gage or other pressure-indicating device located at the boiler or superheater outlets, at the boiler or economizer inlet, and upstream of any shut-off valve used between heat absorption sections. The details of these requirements are contained in PG-60.6.2.

Pressure gages are available in several grades or qualities. The precision or accuracy varies with the quality. Note that when a hydrostatic test is being applied to a boiler or its component parts, a test gage should be used. Precision test gages have an accuracy of one-half of 1 percent. Commercial-quality gages frequently have an accuracy of 2 percent or less and may only be close to correct at the midrange of the dial. Commercial pressure gages operate on the same principles that a precision test gage does, except that materials, construction, assembly, and adjustments of a test gage are more closely controlled and the adjustments more precise.

The Bourdon tube (see photo 8.3) is the component that provides the movement to a pressure gage. This is an alloy tube of oval cross section rolled to a circular shape. One end of the tube is attached to the socket, and the other is plugged. As pressure is introduced into the tube through the socket, the tube tends to straighten out. This causes the tip to move outwards. The tip is attached to the slide on the sector gear so that when the tip of the tube moves, the sector gear is also moved. The wall thickness and alloy used to form the tubes vary with the pressure rating. The travel or movement of the tip also varies. Before adjusting the gage, tension should be applied to the hairspring. Two adjustments are provided: the adjustable link and the sector slide. Lengthening or shortening the sector arm by moving the sector slide changes the movement of the rack and pinion, thus increasing or decreasing the relative rotation of the pointer. Lengthening the sector arm slows it down, and shortening it causes the pointer to move faster. For proper adjustment, the adjustable link should be at a 90° angle to the sector arm when the pressure is at midrange. The length of the adjustable link is set by fixing the length of the link when the pressure is at midrange. Both of these adjustments are determined for each gage. The outer end of the pointer for a precision gage is turned at 90° (i.e., edgewise) to the face of the gage, so it is much easier to more precisely read the pressure. The hand is firmly placed on the pinion shaft with pressure on the gage. The fine adjustment is then made by turning the micrometer adjustment to the exact pressure. This may be turned as many revolutions as required without damage to the hand or gage (see photo 8.4).

A test gage, when being used to apply a hydrostatic pressure to a boiler, should have a range of 1½ times the pressure to be applied. A gage with maximum pressure rating equal to the pressure to be applied should never be used, since control of the pressure is often difficult; the Code allows the actual test pressure to exceed the nominal 1½ times the maximum allowable working pressure by 6 percent (PG-99.1). This overpressure would damage the gage, and the boiler may also be subjected to overpressure due to an incorrect reading on the gage. The dials of test gages should be graduated over the full pressure range of the gage.

High-temperature water boilers are also required to be equipped with a gage that shows the temperature in degrees Fahrenheit. This gage should be located at the outlet, which is the hottest point in the system. See PG-60.6.4.

Gage Glass and Water Columns

The design and application of water columns and gage glasses (see photo 8.5) are covered in PG-60, which contains the requirements for miscella-

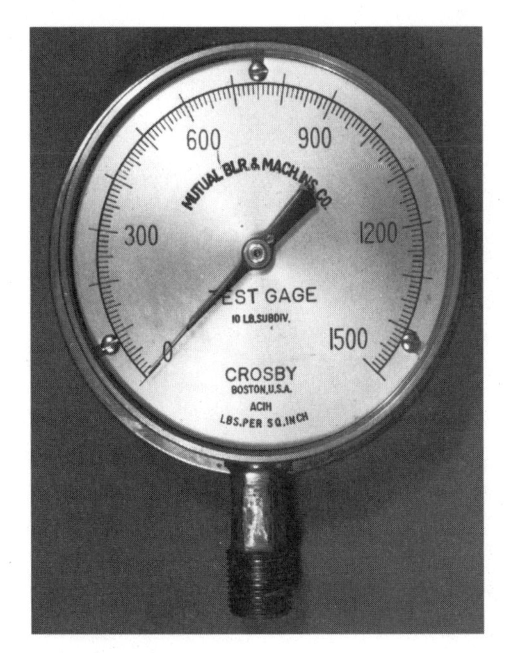

Photo 8.3 A 1500-lb test gage. Note the hand. Photo by the authors.

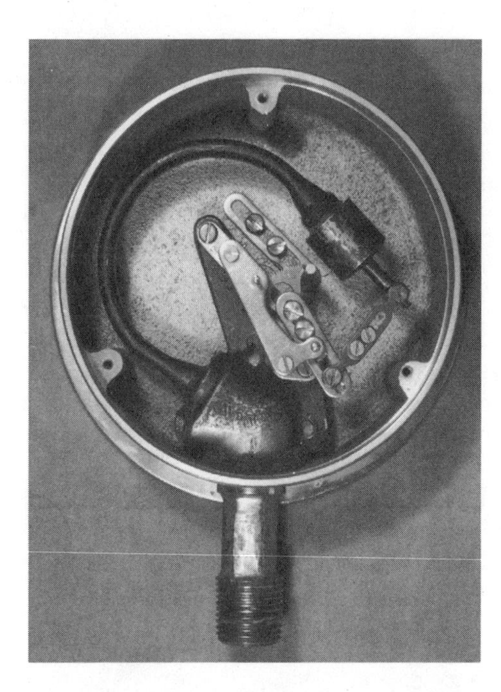

Photo 8.4 Bourdon tube and the movement in the gage. Photo by the authors.

Photo 8.5 A water column with gage glass and gage cocks. Note the chains for blowing down the gage glass and checking the test cocks. Photo by the authors.

neous pipe, valves, and fittings. The piping in this paragraph shall be designed in accordance with the requirements of ANSI/ASME B31.1.

The pressure-rating requirements of gage glasses are given in PG-12. The gage glass must meet a manufacturer's standard. The requirements for water columns are given in PG-11.3 and PW-42, Note 7.

The sizes and pressure ratings of the piping and fittings is given in B31.1, 122.3: Instrument, Control, and Sampling Piping. In general, the piping must be suitable for the maximum safe pressure of the boiler. Again there are special rules for forced-flow steam generators with no fixed water level. See PG-67.4 and B31.1 122.1.1 B.

The requirements for water level indicators are given in PG-60.1.1. The purpose of these requirements is to provide the boiler operator with a means of positively determining the water level in the boiler being operated. Therefore, the requirements for gage glasses do not apply to boilers that do not have a fixed water level, such as high-temperature hot water boilers or forced-flow steam generators (PG-60.1.2).

All steam boilers shall have at least one gage glass, and boilers operating above 400 psi shall have two water gage glasses. (Electric boilers of the electrode type require only one gage glass, per PG-60.1.1 and PEB-13.1.)

On boilers with safety valves set at or above 900 psi, two remote level indicators may be installed in place of one of the water gage glasses (PG-60.1.1), and if the two remote level indicators are in reliable operation, the gage glass may be shut off.

When the water gage glass is not visible from the operating floor, two indirect indicators or two remote indicators shall be provided (PG-60.1.1).

The water gage glass shall be installed so as to indicate a minimum water level at least 2 inches above the minimum level specified by the boiler manufacturer. This minimum level for horizontal firetube boilers shall be 3 inches above the highest level of the tubes, flues, or crown sheets (PG-60.1.4). For locomotive boilers there are similar, size-dependent requirements given in PG-60.1.5.

Connections to the water gage glass shall be a minimum of ½ inch NPS. Gage glasses shall be provided with straight-through shut-off valves to both the top and bottom connection. The glass must be provided with a valved drain at least ¼ inch in diameter (PG-60.1.6). This requirement is assumed also to refer to pipe size (NPS). For a remote level indicator, the piping from the boiler up to the isolation (stop) valve shall be at least ¾ inch pipe size, and from the valve to the level indicator at least ½ inch OD tubing. See PG-60.1.1.

Paragraph 60.1.6 requires the drains on the gage glasses on boilers operating at pressures in excess of 100 psi to be piped to the ash pit or other safe place. It is recommended, although, that the drains of all high pressure boilers be piped to a safe place regardless of the operating pressure.

The lower or water connection to the gage glasses of boilers operating at pressures of 400 psi and above shall be provided with shields or other means to prevent thermal differentials in the shells or heads (PG-60.3.5). Slight temperature differentials often cause cracks to occur—these start at the openings and extend radially into the shell or heads to which they are attached. These areas require careful examination at the annual internal inspections.

Water columns, when used, shall be connected to the boiler by the use of 1-inch pipe. The connections shall be provided with a means of inspection of the internals of the piping. Plugged openings providing easy viewing to ascertain that the piping is clean and unobstructed, are an acceptable means of meeting this requirement. This is commonly accomplished by the use of special cross fittings. The column shall be provided with a valved drain, at least NPS ¾, that discharges to a safe place (PG-60.2.3).

Water columns serve as a means of attaching level-indicating and control equipment to the boiler, such as gage cocks, high- and low-water

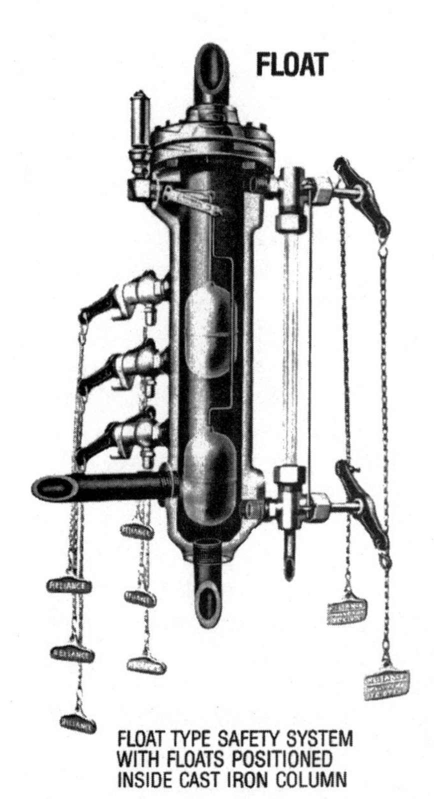

Sketch 8.3 A water column with gage glass, gage cocks, and internal floats for boiler controls and alarms. Courtesy Clark-Reliance Corporation.

alarms using floats or electrodes located within the water column, and the gage glass itself. These units may also provide low- and high-water alarms, feed pump controls, fuel cutoffs, and the like (see sketch 8.3).

The float-type controls are available for use with pressures as high as 900 psi, but they are most frequently used for boilers operating below about 250 psi. Electrical probes may be used to perform the same functions as the float-type units. The various functions are controlled by varying the lengths of the probes fitted into the top of the water column. The probes may activate visual or audible alarms as well as other controls.

Boiler water level indicators are available for direct and remote reading. Direct reading indicators include the tubular, flat, or multiple bull's-eye glasses. These frequently assist the operator by indicating steam as red and water as green. Color readouts are based on a principle of physics: Refraction of light varies for steam and water. By passing a light source through green and red glass filters and then through the gage glass at precisely the proper angle, steam is indicated as red and water as green. Where the gage glass is not readily visible to the operator, the image may be transmitted through fiber optics to a remote readout (see sketch 8.4).

Probe-type water level indicators are available for remote boiler water level indication, and they may also be used to operate alarms, pumps, fuel cutoffs, and the like. The probes are located in a special column and may be spaced as close as 1 inch apart. The probes sense the level of the water and are connected electrically to a control box.

Thermostatically actuated feed water regulators (sketch 8.5) such as the Copes feed water regulator may also be connected to the water column connections. The thermostatic tubes on these units expand and contract with the rise and fall of the water level. This expansion and contraction is mechanically connected to a feed water control valve, causing the valve to open and close to control the water level in the boiler. A major advantage of this type of device is that it needs no outside power for its operation.

Water columns may be fabricated of steel, cast or forged steel, or cast or ductile iron, provided the materials meet the requirements for pressure and temperature as required by PG-60.2.4 and PG-42. The water column and gage glass connections should not have shutoff valves on the connection between the boiler and the column or gage glass unless they meet the requirements of PG-60.3.7. The valves, if used, should be outside screw and yoke or lever lift-gate valves or cock-type valves that visually indicate whether the valves are open or closed. The valves shall also be locked in the open position. The gate-type and cock-type valves provide a clear full opening, making them readily accessible for cleaning and inspection.

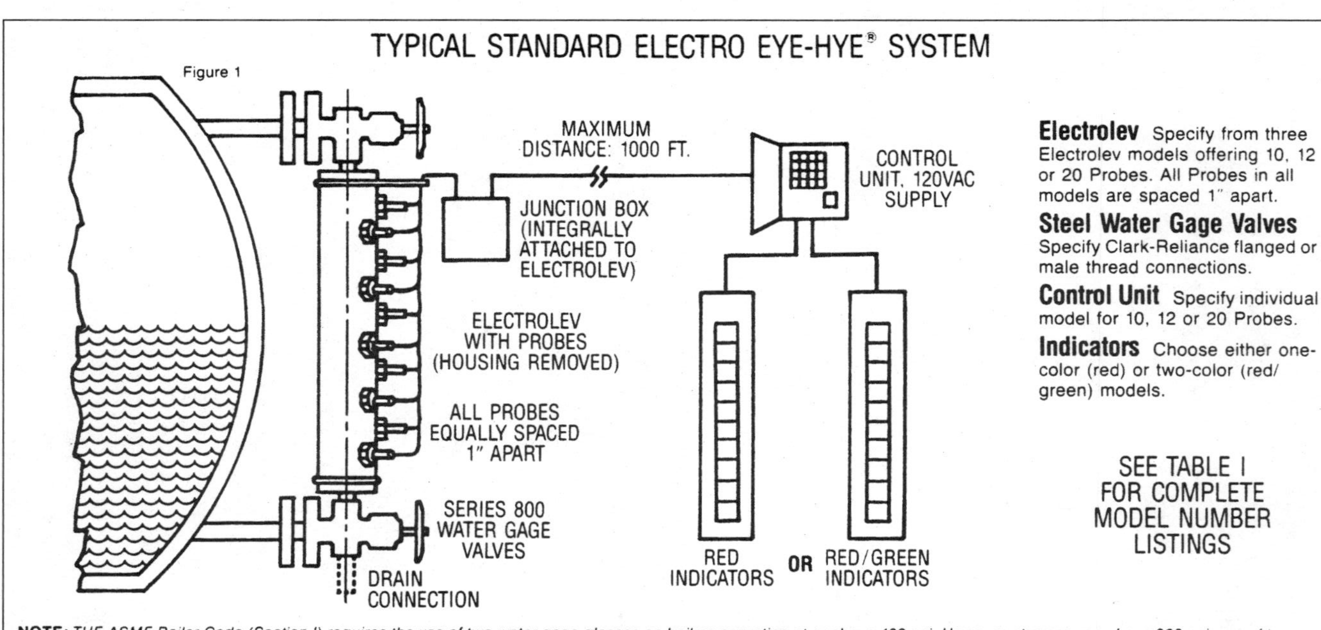

Sketch 8.4 A probe-type remote level indicator system. Courtesy Clark-Reliance Corporation.

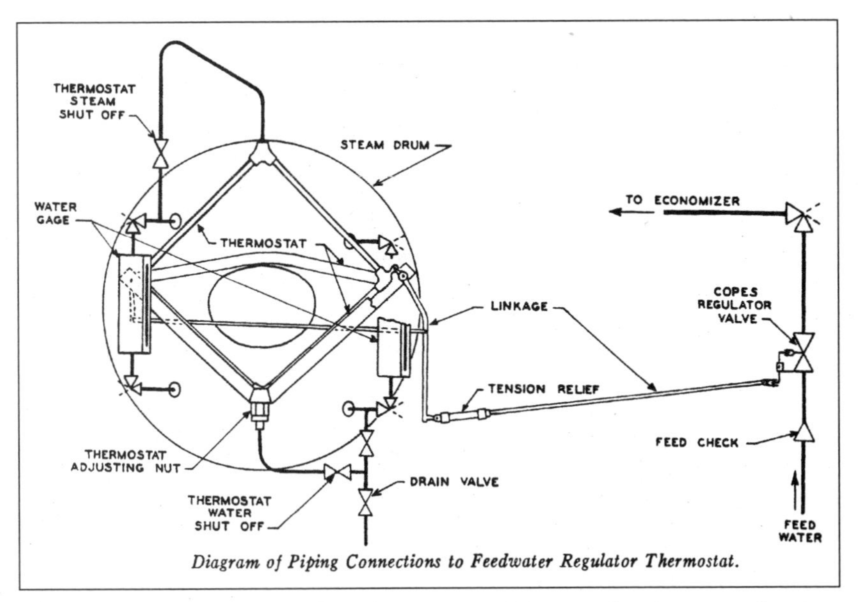

Diagram of Piping Connections to Feedwater Regulator Thermostat.

Sketch 8.5 A feed water regulation system requiring no electric power. Courtesy Copes-Vulcan, Inc.

The only connections permitted to the water column are those that do not require the discharge of steam or water (PG-60.2.6). This eliminates the chance of the other devices causing an incorrect water-level reading.

The importance of being able to reliably determine the water level in a boiler has been demonstrated by the requirement for redundancy in designing and installing the water level indicators from the inception of the first codes. Until recently, gage glasses and gage or try cocks were required on all boilers. The revisions in 1991 of PG-60.4 state the gage cocks are no longer required. The authors, however, recommend the installation of gage cocks on boilers operating at pressures up to and including 400 psi, above which two gage glasses are required.

Steam Piping

Steam, boiler water feed, and blowoff piping are designed in accordance with PG-58 and 59, and B31.1 Part 6 Design Requirements Pertaining to Specific Piping Systems. The limits of Code jurisdiction for piping for drum-type boilers and forced-flow steam generators with no fixed water line are shown in Figures 59.3.1 and 58.3.2, respectively.

The requirements for steam outlet piping are covered by PG-58 and ANSI/ASME B31.1 Part 6, paragraph 122.

Boiler external piping, as it applies to the steam outlet connections, shall include the piping up to and including the steam stop valve for a single boiler installation. This stop valve may be omitted in single-boiler and prime-mover installation that has a throttle valve that is sufficiently designed to withstand the hydrostatic test pressure of the boiler and for which the valve position is clearly indicated.

In boiler installations where two or more boilers with manholes discharge into a common main or header, each boiler or superheater shall be equipped with two stop valves with a drain valve fitted between the stop valves. The main purpose of the two valves with a drain between is the safety of personnel in the empty boiler while the other boiler is steaming. (That is why the requirement applies to boilers with manholes.) The drain should be located so as to permit draining of condensate from the lines and to prevent water hammer, and to indicate whether the line stop valve is leaking when closed and the boiler is off the line. See PG-58.3.1, 58.3.2, and B31.1, paragraphs 122.1.2 and 122.1.7.

It is recommended that the boiler stop valve be a nonreturn type (photo 8.6) and that the line stop valve be an outside screw and yoke type (photo 8.7) located before the line enters the header. The use of nonreturn valves makes it easier to bring a boiler on-line, since the nonreturn valve can be put in the open position before the boiler is up to pressure. Steam from the other boilers will not enter this boiler, since the valve will be held closed by the higher line pressure. As the

Photo 8.6 A boiler stop valve, nonreturn type. Photo by the authors.

Photo 8.7 A line stop valve, gate type, with outside screw and yoke. Photo by the authors.

pressure reaches the pressure in the line, the valve will automatically open and the boiler will be on-line.

When a stop valve of the manually opened type is used, the operator has to watch the pressure carefully as it comes up, and when it is about 3 to 5 psi below the line pressure, the valve should be carefully opened to allow the pressure to equalize. Any condensate that has accumulated above the valve and not drained through the drain between the boiler stop and line stop during warm-up, will drain back into the boiler and not into the line where it could cause water hammer, resulting in damage to the line or equipment being supplied steam.

The design pressure of the piping valves and fittings should be determined in accordance with the requirements of B31.1, paragraph 122, for the pressure and temperature conditions to be encountered.

Boiler Feed Water Piping

The requirements for boiler water feed piping may be divided into three types.

The first type applies to single boiler installations, which require a stop valve and check valve. The stop valve or cock shall be nearest to the boiler.

The positions of the valves may be reversed on a single boiler turbine installation.

The second type applies to plants having two or more boilers fed by a common source. In this case, each boiler should be equipped with a stop valve located near the boiler, then a check valve, and then a globe or regulating valve in the branch to each boiler. When a globe-type valve is used, the feed supply shall come into the valve under the seat. When a combination stop and check valve is used, it should be considered as a stop valve, and a separate check valve is required. Refer to PG-58.3.3 through PG-58.3.5 and Figure PG-58.3.1, as well as B31.1 Part 6, paragraph 122.1.7B.

The third feed system applies to forced-flow steam generators with no fixed steam and water line. The piping requires a stop valve; however, the check valve may be omitted at the boiler, provided a check valve having the required pressure rating is installed at the discharge of each boiler feed pump. This check valve may also be located anywhere in the feed line, between the stop valve at the boiler and the feed pump. See B31.1 Part 6, 122.1.7 B.9.

The feed piping shall be designed for a pressure the lesser of 25 percent or 225 psi above the maximum allowable working pressure of the boiler. The valve of P shall not be taken as less than 100 psi nor less than the pressure required to feed the boiler. See B31.1 Part 6, paragraph 122.1.3 A through C.

Steam boilers and high-temperature water boilers shall have at least one means of feeding water to the boilers while under pressure (PG-61). Boilers having more than 500 square feet of heating surface shall have two or more means of feeding the boiler at a pressure of at least 3 percent above the highest set pressure of any safety valve on the boiler.

Boilers fired with solid fuels other than those fuels fired in suspension, the settings of which may retain sufficient heat to damage the boiler in the event that the feed supply is interrupted, shall have two means of feeding not susceptible to the same interruption. Each should be capable of supplying enough water to prevent damage to the boiler (PG-61.1). This generally means that there must be a tank, reservoir, or other supply of feed water sufficient to prevent overheating of the boiler, in addition to the municipal supply line.

These requirements may be met by having one electric motor-driven pump and one steam-driven turbine or a reciprocating pump or an injector or any combination of these methods of feeding water to the boiler.

Boilers fired by liquids, gas, or solids in suspension may have only one means of supplying water to the boiler, provided the boiler is equipped with a low-water fuel cutoff that activates before the water level goes below that established by the manufacturer or the levels stated in PG-60.1.4 and 60.1.5. The Code is unclear as to whether PG-61.2 requires permit boilers over 500 square feet in heating surface to have a single source of feedwater if they meet the above firing and fuel requirements; it is recommended that all boilers over 500 square feet have at least two independent sources of feedwater.

The minimum size of feed line to boilers having less than 100 square feet of heating surface shall be not less than NPS ½, or not less than NPS ¾ for boilers having more than 100 square feet of heating surface (PG-61.2). PEB-11.2 requires a minimum feed line size of NPS ½ for electric boilers; PMB-11 requires a minimum of NPS ½ for steel or iron pipe; and NPS ¼ for copper or brass.

There shall be a feed water supply capable of introducing water to high-temperature water boilers at the operating pressure. See PG-61.4. There shall also be a means of feeding water into a forced-flow steam generator with no fixed water level while it is operating at the maximum designed pressure and capacity of the boiler with the maximum allowable working pressure at the superheater outlet. See PG-61.5.

These requirements can be condensed into one sentence: Every boiler shall have at least one means of feeding water to it at the maximum allowable working pressure, having sufficient capacity to supply all the steam or water being discharged from the boiler; and boilers having more than 500 square feet of heating surface or burning solid fuel shall have two such means, independent of each other.

Blowoff, Blowdown, and Drain Piping

The requirements for the design and use of boiler external piping used for blowing off, blowing down, and draining vary with the use of the particular connection and valve or valves. One needs to read carefully all paragraphs in both Section I and B31.1 to understand how the differentiation is made.

Blowoff and blowdown connections are provided to blow water out of the boiler while the boiler is under pressure. To differentiate between them, one must consider the specific uses of each.

A blowoff line is used intermittently to remove sediment, reduce the concentration of solids, or rapidly lower the water level in the boiler.

All boilers shall have a blowoff line connected to the lowest point in the boiler, so that it can be opened intermittently to remove sediment and/or dissolved solids during operation. It is also used to lower the water level and, during shutdown, to drain the water from the boiler.

Water wall headers, water screens and the like that do not drain back into the boiler shall be equipped with blowoffs or drain connections. See PG-59.3.3 and 59.4.1. When blowoffs are exposed to direct furnace heat, they shall be adequately protected by insulation, firebrick piers, and so on (PG-59.3.7). An example is the blowoff connection on an HRT boiler.

Blowdown lines or connections are in continuous use to control the dissolved solids in the boiler to reduce scale formation, priming, and carryover. See Figure PG-58.3.1 and PG-58.3.7.

As evidenced by the use requirements, blowoff connections and valves are subject to thermal shock and, in many instances, to impact or water hammer. Therefore, the design requirements include the use of a higher value of pressure *P* in the formulae in Part 2 of B31.1, paragraph 104.

The Code jurisdiction for blowoff piping extends from the boiler to and including the first valve on boilers with a maximum working pressure of less than 100 psi; on forced-flow steam generators; on traction and portable boilers; and on electric boilers with a maximum capacity of 100 gallons. All other boilers are required to have two valves in the blowoff line (PG-58.3.6).

When the drain connection on a high-temperature water boiler is not intended to be used as a blowoff, it may have one stop valve; otherwise, two valves are required. See PG-58.3.6 and PG-58.3.7.

Blowoff piping up to and including the required valve or valves shall be designed for a pressure of 25 percent or 225 psi higher than the maximum allowable working pressure of the boiler. See B31.1 122.1.4 A1 through A4. The blowoff connection shall be at least the size of the boiler connection, but not more than 2½ inches. The minimum size of blowoff connections shall be 1 inch NPS except for miniature boilers. The minimum size of the blowoff for a miniature boiler is ½ inch (PMB-12). For an electric boiler with input power of 200 kW or less, the blowoff may be ¾ inch; otherwise, it shall be 1 inch NPS (PEB-12).

All piping used in the blowoff system shall be steel. Galvanized pipe and fittings shall not be used in the blowoff lines. Bronze, cast iron, and ductile or malleable iron fittings as well as steel fittings may be used in the blowoff system for boilers having a maximum allowable working pressure not exceeding 100 psi. Refer to B31.1 Part 6, 122.1.4 A to A4. All boilers, except traction and portable steam boilers but including electric boilers with water capacities of over 100 gallons and operating at over 100 psi, shall have two bottom blowoff valves. Both may be slow opening, or one may be a cock or quick-opening valve. See B31.1, 122.1.7, C1 through C12.

Traction, portable forced circulation, and electric boilers not covered above and steam boilers operating below 100 psi shall have at least one slow-opening valve.

Surface blowoff connections shall not exceed 2½ inches and shall be connected through a bushing or flanged connection as shown in Figure PG-59.1. A surface blowoff is used intermittently to remove surface contaminants such as oil from the surface of the boiler water. All surface blowoffs are intended to be used when the boiler is under pressure, so all surface blowoffs shall have two valves under the conditions

Photo 8.8 Blowoff valves. The lever-operated valve on the left is nearest the boiler and is fast opening. The other valve is slow opening. Photo by the authors.

where two would be required for other blowoff connections (photo 8.8). When drain lines are not used or intended for blowoff purposes, they may have only one valve per PG-58.3.7.

A boiler should never be blown down until the water level in the boiler has been verified by blowing down the gage glass and water column. The procedures for blowing down the different types of boilers are given hereafter for firetube boilers and tank-type boilers generally and for steam drums in watertube boilers.

The requirements for the location of the quick-opening valve next to the boiler and the slow-opening valve away from the boiler have been determined by in-service experience. Since blowing down a boiler requires slowly opening the blowoff valve to prevent water hammer and extreme thermal stresses, the valve used for blowing down is most vulnerable to damage by erosion. The quick-opening valve should be opened slightly; the pressure between the valves is equalized very quickly, and then the valve is opened fully. This valve is therefore not subjected to severe erosion. The slow-opening valve is then opened slowly and reclosed slowly when a sufficient amount of water has been blown out of the boiler. If it is not possible to observe the gage glass from the location of the blowdown valves, the level is checked after the valves are closed, and if additional water must be blown down, the procedure is repeated. It is vital that the operator never leave the blowdown valves while they are open for any purpose whatever; an operator can become distracted by a phone call or some other cause, and the boiler may be blown dry. After the completion of the blowdown, the slow-opening valve is closed; then the fast-opening valve is closed; and then the slow-opening valve is opened again slightly to drain the line between the valves and to test the tightness of the other valve.

The process of blowing down waterwall headers is more demanding and risky because of the lesser volume of water in the system and the danger of reversing the circulation. The manufacturer's instructions should be followed, but the following general points are given: The load on the boiler should be reduced to between 25 and 50 percent of its rated capacity before opening the blowoff valves as described earlier. It should be emphasized that the reason for blowing down waterwall headers is to remove the sediment that can otherwise build up in areas of poor circulation. Several short blows are more effective for this purpose than a single long one, and there is less chance of reversing the circulation with consequent tube ruptures in the waterwalls.

As indicated above, the lines for blowing down the gage glass and water column are considered to be drains rather than blowoff lines. Other drain lines are discussed in PG-59.4. Drains are required for piping, superheaters, waterwalls, and generally all areas where water can accumulate. In particular, at least one drain is required on a high-temperature water boiler and a superheater (PG-59.4.1 and 59.4.2). It is again noted that if a drain line is to be used as a blowoff line as well, it must have two valves and meet all the design requirements of B31.1, paragraphs 122.1.4, 122.1.7 (g), and 122.2.

Sootblowers

PG-68.5 states: "Sootblowers may be attached to the same outlet for the superheater or reheater that is used for the safety valves." Sootblowers are devices used intermittently to remove accumulations of soot and fly ash from the fire side to promote heat transfer and prevent plugging up of the gas passages. There are three basic methods of blowing soot from boilers. The first and most primitive is the use of manually operated steam lances inserted through holes in the refractory into the critical areas of the boiler.

The second method is the use of a fixed sootblower, which may be manually operated or motor-driven. The blowing element is in a fixed position in the boiler and may be rotating or nonrotating. The nonrotating type is so positioned and directed that it clears the target areas without motion of the element. On rotating units (photo 8.9), there are several nozzles at specific locations along the length, depending on the tube spacing and geometry of the particular boiler. The nozzles are designed to blow through a lane as the sootblower element rotates

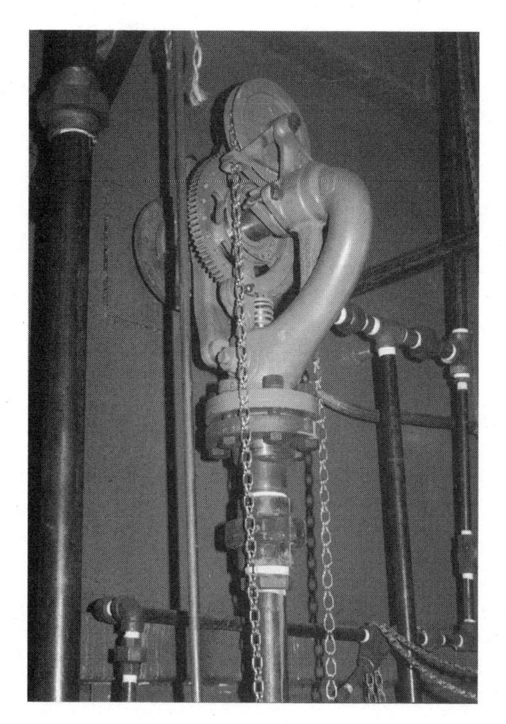

Photo 8.9 A fixed-type sootblower, manually operated, with rotating element. Photo by the authors.

through its arc in order to remove soot, slag, and the like from the tube bundles. To conserve steam, the nozzles are controlled by a cam and valve arrangement that starts and stops the steam flow through the individual nozzles at the appropriate points in the element's rotation. It is important that the sootblowers be properly adjusted and maintained to prevent steam from blowing directly on the tubes. The steam pressure to the sootblowers is also adjusted to match the fuel being burned and the boiler design. Too high a pressure or moisture in the steam may cause erosion of the tubes.

Boilers fired by oil or pulverized fuel should be operating at sufficient capacity to prevent concentrations of particles from reaching levels high enough to cause an explosion. Generally the sootblowers should not be operated when the boiler is running at less than 50 percent of rated capacity.

In between the times of sootblower operation, a flow of air must be maintained through the blowers to prevent them from overheating.

The third sootblowing system (sketch 8.6) incorporates a lance that is attached to a carriage that moves on rails, carrying the lance forward into the boiler. The lance is fitted with one or more nozzles, and as it advances it also rotates. The nozzles therefore travel helical paths. When the lance reaches the extreme end of its travel, it is indexed, such

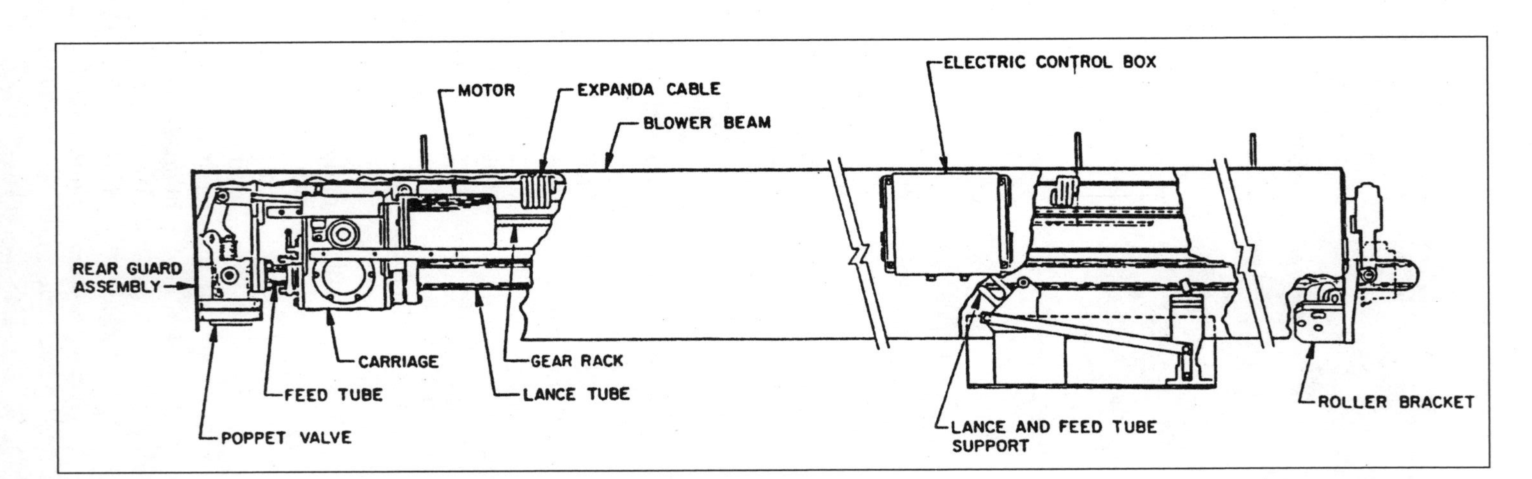

Sketch 8.6 A retractable-type sootblower. Courtesy Diamond Power Specialty Company.

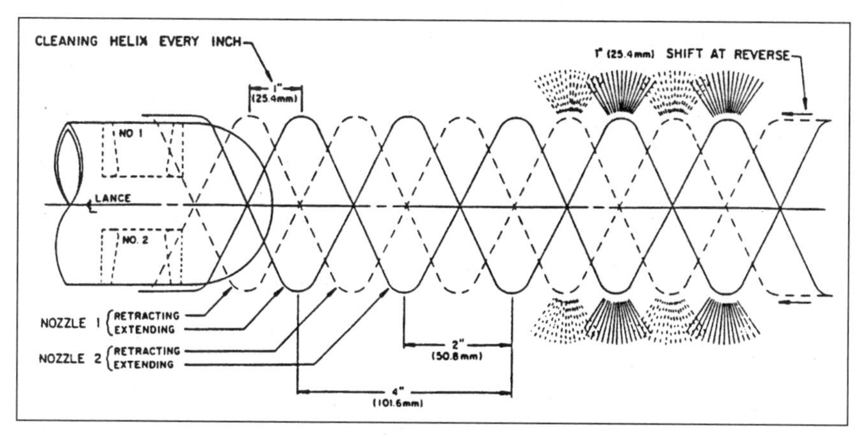

Sketch 8.7 Cleaning pattern for a retractable-type sootblower. Courtesy Diamond Power Specialty Company.

that each nozzle travels a different helical path as the lance is retracted from the boiler. As the nozzle approaches the furnace wall on its retract cycle, the blowing medium is shut off. The carriage continues to the fully retracted position. The design must incorporate provisions to prevent overheating of the sootblower and to prevent furnace leaks, whether the furnace is at positive or negative pressure.

This third type of sootblower is generally found on large boilers with several to many sootblowers. Again, sootblowing must be done at a sufficiently high boiler capacity to clear the boiler of the materials being cleaned from the boiler surfaces. Blowing on large boilers should start nearest the furnace and progress in the direction of combustion gas flow, through the gas passes. In large furnaces, the danger of a furnace explosion is less than in smaller ones, but normal precautions should still be taken. This requires an adequate gas flow through the boiler, as indicated above.

Fusible Plugs

Fusible plugs are not required for steam boilers in Section I. The rules for fusible plugs are given in the Appendix, in A-19 through A-21. Fusible plugs were widely used in the era of hand-fired locomotive, vertical firetube, and HRT boilers firing solid fuel. In the vertical-tube boilers, the fusible plug was in a tube wall at the waterline. In HRTs, the fusible plug was located in the rear head, above the tubes. In the locomotive boilers, the fusible plug was in the crown sheet, right above the fire. The plug was intended to melt and admit steam to the furnace to smother the fires in order to prevent overheating damage to the boilers.

It is considered highly questionable whether a fusible plug ever did perform as the designer intended. The draft in the boiler is generally sufficient to clear the steam from the furnace, and the plug only serves to increase the water loss from an already overheated boiler. Further, there is clearly no advantage to using a fusible plug when firing liquid, gaseous, or pulverized fuels; in these cases, the fire is best extinguished by the low-water fuel cutoff or by the fireman alerted by a low-water alarm. Indeed, fusible plugs serve no useful purpose in today's steam boilers, and their use is not recommended. In existing boilers, the fusible plug can be replaced by a steel plug to close the opening.

Chapter

9

Maxima, Minima, Et Cetera

The Code contains many requirements in the form of stated maximum or minimum criteria (e.g., maximum pressure on gray cast iron, minimum thickness of a plate to be stayed), and a number of other rules that must first be discovered, then remembered. This chapter is intended to assist the reader in navigating the Code for little known rules. In many cases, there are short rules, accompanied by somewhat longer qualifiers. The information tabulated below gives the subject, the paragraph containing the applicable rule, and a short version of the rule itself. It must be emphasized, however, that not all of the qualifiers are included in these brief summaries; to do any actual boiler work, it will be necessary to read the paragraphs themselves for the correct application.

Maximum Criteria

Subject	Location	Summary
Adjustment on a safety valve spring	PG-72.3	5%
Blowdown of a safety valve	PG-72.1	4% of set pressure
Carbon content of a steel to be welded	PW-5.2	0.35%
Compressed thickness of manhole or handhole gasket	PG-44.3	¼″
Deviation of a shell from circularity, external pressure	PG-80.2	1% up to 24″, then "e" from Figure PG-80
Deviation of a shell from circularity, internal pressure	PG-80.1	1%
Diameter of a miniature boiler	PMB-2	16″ ID
Diameter of a part calculated by the tube formula	PWT-10.2	5″
Diameter of a seamless pipe threaded into a drum	PWT-9.2	NPS 1½

Subject	Location	Summary
Flanged-in opening not requiring added thickness	PG-29.3	6″
Geometric unsharpness of a radiograph	PW-11.2.1	0.07″
Heating surface of a miniature boiler	PMB-2	20 ft^2
MAWP of a miniature boiler	PMB-2	100 psig
MAWP on ASME/ANSI B16.5 pipe flanges	Table A-361	See table
Metal temperature in the hydrostatic test	PG-99.2	120°F
Pipe size for socket welds	PW-41.5.1	NPS 3
Pitch of threaded stays	PG-46.5	8½″
Pitch of welded stays	PG-46.5	15 diameters
Popping point tolerance of safety valve	PG-72.2	See PG-72.2
Pressure for cast iron handhole plates	PG-44.2	250 psig
Pressure in a boiler with safety valves blowing	PMB-15	106% of MAWP or 6% over MAWP
	PG-67.2	106% of MAWP or 6% over MAWP
Pressure in a hydrostatic test	PG-99.1	6% above test pressure
Pressure on nodular iron boiler parts	PG-8.3	350 psig
Pressure on threaded joint	PG-39.5.2	See table
Projection of welded stays on fire side	PW-19.3	⅜″
Projection of a watertube before expanding	PWT-11.1	¾″
Punched hole for stay in plate ≤ 5⁄16″ thick	PG-82	⅛″ less than stay diameter
Punched hole for stay in plate > 5⁄16″ thick	PG-82	¼″ less than stay diameter
Punched tube hole	PG-79	½″ less than tube OD
Reinforcement on the face of a weld	PW-35.1	See table, PW-35.1
Root concavity of a weld	PW-41.2.2	3⁄32″ or 20%
Safety valve attached by threaded connection	PG-67.6	3″
Size of a surface blowoff line	PG-59.3.2	NPS 2½
Size of a blowoff connection	PG-59.3.5	NPS 2½
Size of welded pipe connection not requiring inspection	PW-41.3	NPS ½
Stud size for stud welds	PW-27.2	1″ diameter
Tell-tale hole diameter in a reinforcing pad	PW-15.3	¼″ pipe tap
Temperature for cast iron handhole plates and other boiler parts	PG-44.2 PG-8.2.2	450°F
Temperature for nodular iron boiler parts	PG-8.3	450°F
Temperature for a threaded boiler connection	PG-39.5.2	925°F
Trough of a stayed box header	PWT-12.1	0.9P
Tube attached by expanding	PG-39.6	6″ OD
Tube size for socket welds	PW-41.5.1	3½″ OD
Uncalculated, uncompensated opening	PG-32.1.3.1.1	NPS 2
Undercut of a weld	PW-35.1	lesser of 1⁄32″ or 10%
Undertolerance of a plate	PG-16.4	0.01″
Volume of a miniature boiler	PMB-2	5 ft^3
Weld reinforcement	PW-35.1	3⁄32″ to 3⁄16″
Width of a welded stayed water leg	PWT-12.2	4″

Minimum Criteria

Subject	Location	Summary
Access opening in smoke hood of vertical firetube boiler	PG-42	6″ × 8″
Bearing width of a manhole gasket	PG-44.3	11⁄16″
Blowdown of a safety valve	PG-72.1	2% of set pressure
Clearance under the furnace of wet-bottom firetube boilers	PFT-46.6	12″
Connections on the gage glass	PG-60.1.6	NPS ½
Connection for the pressure gage, ferrous	PG-60.6.1	½″ ID
Connection for the pressure gage, nonferrous	PG-60.6.1	NPS ¼
Connections to the remote level indicator	PG-60.3.4	NPS ¾
Connection for the test gage	PG-60.6.3	NPS ¼
Connections to the water column	PG-60.3.4	NPS 1
Depth of insertion of a tube or pipe in a socket weld	PW-41.5.2	¼″
Design temperature of a tube receiving heat	PG-27.4, Note 2	700°F
Diameter of a tell-tale hole in a staybolt	PG-47.1	3⁄16″
Distance between centers of reinforced openings	PG-38.2	1⅓ times the average diameter
Distance between through stays in the bottom of an HRT boiler	PFT-24.4	10″
Drain cock for the gage glass	PG-60.1.6	¼″
Drain for a safety valve body ≤ 2½ NPS	PG-73.1.5	NPS ¼
Drain for a safety valve body > 2½ NPS	PG-73.1.5	NPS ⅜
Drain cock for the water column	PG-60.2.3	NPS ¾
Extension of staybolt tell-tale hole inside edge of plate	PG-47.1	½″
Fillet weld size on a diagonal stay	PW-19.4	⅜″
Flare of a watertube	PWT-11.1	⅛″ larger than the diameter
Hole in splash plate or collecting pipe for safety valves	PG-71.7	¼″
Hydrostatic test pressure of a boiler	PG-99	1.5 MAWP
Hydrostatic test pressure of a miniature boiler	PMB-21	3 MAWP
Material to be machined from a sheared nozzle end	PG-76.2	⅛″
Metal temperature during the hydrostatic test	PG-99	70°F
Number of gage glasses required for boilers operating over 400 psig	PG-60.1.1	2
Number of gage glasses required for boilers with drum safety valves set ≥900 psig	PG-60.1.1	1, plus 2 remote level indicators
Number of handholes or washout plugs in a Scotch boiler	PFT-43.4	4
Number of handholes or washout plugs in a vertical firetube boiler	PFT-43.5	
>24″ OD		4
<24″ OD		3
Number of threads for a threaded connection	Table PG-39	See table
Opening in firetube boiler shell for safety valves	PFT-44	See table, PFT-44

Subject	Location	Summary
Pressure of feed pumps	PG-61.1	3% over highest safety valve setting
Projection of a watertube before expanding	PWT-11.1	¼″
Range of pressure gage	PG-60.6.1	1½ times the safety valve setting
Relieving capacity of reheater safety valve	PG-68.4	15% of total
Safety valve capacity on an electric boiler	PG-68.3	3½ lb/hr/kW
Size of a blowoff connection	PG-59.3.5	NPS 1, heating surface > 100 ft^2 or NPS ¾, heating surface ≤ 100 ft^2
Size of a blowoff connection in a miniature boiler	PMB-12	NPS ½
Size of feed lines	PG-61.3	NPS ½ and NPS ¾ heating surface ≤ 100 ft^2 and >100 ft^2
Size of a firing or access door in a boiler setting	PWT-15	12″ × 16″
Size of a handhole	PG-44.1	2¾″ × 3½″
Size of letters stamped on boilers	PG-106.4	5/16″
Size of a manhole	PG-44.1	12″ × 16″ or 15″ diameter
Size of a safety valve on a miniature boiler	PMB-15	NPS ½
Size of a syphon	PG-60.6.1	¼″ ID
Size of washout plugs	PFT-43.1	1½″
Size of washout plugs in a miniature boiler	PMB-10.1	1″
Thickness of Adamson furnace	PFT-16	5/16″
Thickness of a boiler plate	PG-16.3	¼″
Thickness of a nozzle neck	PG-43	not less than the smaller of the thickness of shell or head; or the thickness of standard wall pipe if the nozzle is pipe; or tube having 600 psig as its MAWP if the nozzle is tube
Thickness of pipe over 5″, used as a shell	PG-16.3	¼″
Thickness of a plate to be stayed	PG-16.3	5/16″
Thickness of plate in an electric boiler	PEB-5.2	3/16″
Thickness of plate in a miniature boiler	PMB-5.2	¼″
Thickness of ring-reinforced furnace	PFT-17.5	5/16″
Thickness of seamless shell in a miniature boiler	PMB-5.2	3/16″
Thickness of shells and domes of firetube boilers	Table PFT-9.1	See table
Thickness of stayed dished head	PG-30.1.2	⅞″
Thickness of tubesheets of firetube boilers	Table PFT-9.2	See table
Thickness of a tubesheet with tubes rolled in, in a miniature boiler	PMB-5.2	5/16″
Thickness of tube with fusible plug	PWT-9.3	0.22″
Time of retention of radiographs	PW-51.4	5 years
Time of retention of report of ultrasonic examination of welds	PW-52.2	5 years

Subject	Location	Summary
Water depth over tubes of horizontal firetube boiler at lowest gage glass reading	PG-60.1.4 and PFT-47	3″
Water depth above lowest permissible level at lowest gage glass reading	PG-60.1.1	2″

Miscellaneous Criteria

Alignment tolerances for butt welded joints are given in table in PW-35.1.

Backing strips must be removed from longitudinal welds but may be left in place on circumferential welds (PW-35.3).

Boiler external piping may be fabricated by either an S stampholder or a PP stampholder (PG-109.1) and may be installed by welding by an S, PP, or A stampholder.

Design temperature of firetubes is 700°F per PFT-50.1 and PG-27.4, Note 2.

Design temperature of furnaces is water temperature + 100°F per PFT-50.1 and PFT-17.7.

The diameter of a hole for checking thickness of corrugated furnaces is ⅜ inch (PFT-18.2).

Exposure of a part to furnace gases occurs if the furnace gas temperature exceeds 850°F (PW-41.1.4.2).

A flanged-in manhole reduces the area to be stayed by 100 in^2 (PFT-27.9).

Heating surface to be measured on the side receiving heat (PG-101.1.1).

Nondestructive testing methods are as follows: radiography (RT); ultrasonic testing (UT); magnetic particle inspection (MT); and liquid penetrant inspection (PT).

Nondestructive testing personnel certifications are issued per employer's written practice using SNT-TC-1A as a guideline (PW-51.5, PW-52.4, and PG-25.2.4). The 95 Code references the 1984 Edition of SNT-TC-1A (A-360).

Plate identification must be maintained on the finished boiler (PG-77.1). See PG-77.2, 3, and 4 for the procedures for transferring markings on plates.

Preheat requirements are found in the Appendix, A-100.

The pressure gage connection required to permit installation of a gage to verify the accuracy of the boiler pressure gage must be valved such that the test gage can be installed, used, and removed while the boiler is steaming (PG-60.6.3).

The radiant heat zone is the first five rows of tubes from the furnace (PW-41.1.4.1).

Radiographic acceptance standards are as follows: No cracks, lack of penetration, or lack of fusion per PW-51.3.1; elongated indications per PW-51.3.2 (¼ inch for t up to ¾ inch; ⅓t up to t = 2¼ inches; and ¾ inch for t > 2¼ inches; and ¾ inch for t > 2¼ inches); and rounded indications (porosity) per A-250 Table 1 and the charts Figures 1 and 2, Figures 3.1, 3.2, 3.3, 3.4, 3.5, and 3.6.

Radiographic requirements are, in general, that all longitudinal and circumferential butt welded joints must be radiographed. There are, however, exceptions to this rule. These exceptions include:

1. Specific exceptions found in other parts of the Code
2. Circumferential welds ≤ NPS 10 or 1⅛-inch wall thickness
3. Pipes, tubes, and headers as described in PW-41

Riveted construction may still be performed using the 1971 edition of Section I (Part PR).

Tack welds that remain in the finished weld must be made by qualified welders. Tack welds that are removed must nonetheless be made by a qualified procedure (PW-31.3).

Temperature gages required on high-temperature water boilers must read in degrees Farenheit (PG-60.6.4).

Ultrasonic acceptance standards are as follows: no cracks, lack of fusion, or lack of penetration (PW-52.3.1); and indications exceeding reference level and length of ¼ inch for $t \leq$ ¾ inch, ⅓t for ¾ < $t \leq$ 2¼ inches, and ¾ inch for t > 2¼ inches (PW-52.3.2).

Ultrasonic testing requirements are discussed in the following sections: UT welds that would otherwise be radiographed if the geometric unsharpness exceeds 0.07 inch, PW-11.2.1; UT electroslag welds, PW-11.2.2; and UT inertia and friction welds requiring radiography, PW-11.2.3.

Welded joints from one side only can be considered equal to double-welded joints "by providing means for accomplishing complete penetration" (PW-9.1).

Chapter

10

Obtaining S and R Stamps

The preceding chapters have concerned themselves primarily with the specific design and material requirements of the power boiler section of the ASME Boiler and Pressure Vessel Code. Code construction, however, involves more than simply buying the Code book and following its design rules. The ASME Code was designed to be enacted into law in order to provide a consistent set of rules for use throughout the United States and Canada. A state or other Jurisdiction can require that boilers to be used in that state be constructed in accordance with the Code and that they be stamped. A stamp is a symbol applied to the boiler or pressure vessel to demonstrate that the vessel meets all requirements of the Code. In order to have the right to apply a Code stamp to a vessel, an organization must have a Certificate of Authorization. The Certificate of Authorization is issued by the Boiler & Pressure Vessel Committee of the American Society of Mechanical Engineers (see PG-105.2). The stamps themselves are also issued by the ASME; they remain the property of the ASME and must be surrendered when an organization ceases to be authorized to perform Code construction for whatever reason. Holders of a Certificate of Authorization can build and stamp boilers, which will then be acceptable in all Jurisdictions that have adopted the ASME Code as a matter of law. The stampholder (also known as the holder of the Certificate of Authorization, and, henceforth, as the Manufacturer) need not be located in the Jurisdiction where the boiler is to be used, or even in any Jurisdiction at all. The Manufacturer can be located anywhere in the world.

Under the ASME Code, the Manufacturer has the responsibility for all Code compliance, from design through materials and fabrication. The Code committee writes the rules, and the manufacturer follows them. There is a third element to the system that relies on independent third-party inspection and verification of the boiler construction to assure that in fact the applicable provisions of the Code have been complied with. The third-party inspection is carried out by Authorized Inspectors (AIs) who must be in the employ of an Authorized Inspection Agency. An Authorized Inspection Agency may be a Jurisdiction or an insurance company that writes boiler and pressure vessel insurance. The Authorized Inspector must be an employee of an Authorized Inspection Agency and needs to have passed a written examination under the rules of any state of the United States or any province of Canada that has adopted the Code as law (PG-91). In short, then, an AI must be employed as an AI by an AIA, and must be qualified by the rules of some Jurisdiction (PG-91). In practice, however, the AIs are required to be licensed (commissioned) in accordance with the requirements of the National Board of Boiler and Pressure Vessel Inspectors.

The definitions and rules of the National Board are contained in the National Board Inspection Code (known as the NBIC), ANSI/NB-23; and in the *Rules and Regulations* (formerly the *Bylaws*), NB-215. The National Board is defined in the Preamble to the *Rules and Regulations:* "The National Board of Boiler and Pressure Vessel Inspectors is an Organization comprised of Chief Inspectors of states and cities of the United States and Provinces of Canada and is organized for the purpose of promoting greater safety . . . among the jurisdictional authorities responsible for the administration and enforcement . . . of the ASME Boiler and Pressure Vessel Code." The Jurisdictions represented on the National Board must have adopted at least Section I of the Code as law. The NBIC and the *Rules and Regulations* make a distinction between the National Board Commissioned Inspector, who has passed the National Board examination and has been issued a certificate of competency to perform inservice repair and alteration inspections and who is regularly employed by an Authorized Inspection Agency and holds a National Board commission; and the Authorized Inspector, who must hold a Certificate of Competency and a National Board commission with an A or B endorsement. The A endorsement allows an Inspector to perform inspections of boilers and pressure vessels under construction. The B endorsement is the Inspector Supervisor endorsement; all A-endorsement Inspectors must be supervised by a B-endorsement Inspector Supervisor. Further, if a B endorsement holder does perform shop inspection, he must be supervised by another B endorsement holder. The duties of AIAs, AI supervisors, and AIs are given in the *Rules and Regulations.* The duties of the National Board commissioned Inspector are given in the NBIC; the specific requirements regarding repairs and alterations are given in R-301.3 of the NBIC.

The National Board examination is given quarterly on the first Wednesday and Thursday of the months of March, June, September, and December. The first day of the examination is a full-day, open-book exam consisting of ten mathematical problems. The books allowed are generally the ASME Code, Sections I and VIII, Division 1, B31.1 (Power Piping), and Section II, Part D. Recent exams have taken approximately half their problems from Section I and B31.1 (i.e., boiler external piping) and approximately half from Section VIII, Division I. The second day is a half-day, thirty-question, closed-book exam. Ten of the questions involve inservice inspection (i.e., the NBIC), and 20 questions involve the ASME Code. These latter questions may involve Sections I, VIII, Division 1, IV, V, and IX. The minimum passing score on the combined examination is 70 percent.

The Manufacturer has a relationship, then, with ASME, with an Authorized Inspection Agency, with an Authorized Inspector, with the Jurisdiction, and indirectly with the National Board. It must be emphasized that the ASME Code applies to new construction only. The repair of boilers and pressure vessels is not within the scope of the ASME Code. Therefore, an analogous set of relationships has been developed associated with the R stamp. Certificates of Authorization to repair and/or alter boilers and pressure vessels are issued by the National Board of Boiler and Pressure Vessel Inspectors, and the R stamps are loaned by the National Board to the Repair Organizations. The Repair Organization must have a contract with an Authorized Inspection Agency to perform inspection of repairs and alterations. The actual inspection work will be performed by an inspector employed by the Authorized Inspection Agency; the Inspector must hold a National Board commission but is not required to hold an A endorsement. (Until recently, a holder of a manufacturing stamp such as the S stamp could perform repairs on vessels within the scope of his authorization to manufacture. Now, however, a repair Certificate of Authorization and an R stamp from the National Board are required, even for organizations holding certificates of authorization from the ASME.)

Section I of the Code encompasses the use of six stamps. They are:

1. The S stamp for manufacturing power boilers.
2. The M stamp for manufacturing miniature boilers.
3. The E stamp for manufacturing electric boilers.
4. The A stamp for boiler assembly.
5. The V stamp for safety valve manufacture.
6. The PP stamp for power piping (i.e., boiler external piping).

There are, of course, many other stamps associated with other Code sections. The process for applying for any of those stamps would be similar to that discussed here. It should be noted, however, that there is a difference in the process of obtaining a V (safety valve manufacturer) Certificate of Authorization and stamp, as compared with the other stamps associated with Section I. The valve stamp will be discussed separately, after the others.

The first step in applying for the Certificate of Authorization is the decision as to what scope of manufacturing activity is contemplated. The S stamp is the most general and allows the widest range of production compared with, for example, the M stamp. Further, it must be realized that not all Jurisdictions have adopted all parts of the Code. For example, Section I indicates in the Preamble and in PEB-3 that the boiler pressure vessel of an E-stamped boiler may be constructed to Section VIII, Division 1, but not all Jurisdictions accept this construction. Information about the boiler and pressure vessel laws in the various Jurisdictions is contained in a publication of the Uniform Boiler and Pressure Vessel Laws Society, Inc., called the *Synopsis.* This looseleaf book may be purchased from the Society by writing to:

Uniform Boiler and Pressure Vessel Laws Society, Inc.
308 N. Evergreen Road, Suite 240
Louisville, KY 40243
(502) 244-6029
FAX (502) 244-6030.

The *Synopsis* is published every two years, and looseleaf updates are issued semiannually. The list of jurisdictional authorities with addresses and phone numbers at the end of this chapter is taken from the 1994 edition, July 1995 update.

When the type of stamp desired has been selected, it is necessary for the would-be manufacturer to have an inspection contract with an Authorized Inspection Agency (PG-105.3). (A list of Authorized Inspection Agencies offering inspection services for new construction is included at the end of this chapter.) The Authorized Inspector from the agency will generally assist the Manufacturer in his application preparations. The Manufacturer must write a quality control manual and implement a quality control system in accordance with the requirements outlined in A-300 (of the Appendix to Section I). When the Manufacturer has his inspection contract and quality control system in place, he will write to the ASME to request a joint review of the Manufacturer's facilities. The address is:

The American Society of Mechanical Engineers
Accreditation/Certification Processing Dept.
22 Law Drive
P.O. Box 2900
Fairfield, NJ 07007-2900.

In response, the Manufacturer will receive a three-page letter explaining the application process (Form letter A1.8); a copy of the ASME application for accreditation; a copy of Form A4.7 Form D dealing with the logistics of the review; a list of Authorized Inspection Agencies offering inspection services for new construction (insurance companies); a list of accepted safety valve testing labs; a Code book order form (showing the volumes required for the various stamps); and information on payment of application fees. At this writing, the application fee for nearly all the ASME Certificates of Authorization, including the six associated with Section I, is $1500 per certificate. The actual stamps are priced at $150. In addition, the Manufacturer pays the cost of the review, including the expenses of the review team.

After the application and fees are received, the ASME will request verification of the inspection contract from the Authorized Inspection Agency. When such verification is received, the ASME notifies the Jurisdiction that it and a representative of the AIA may conduct the joint review of the Manufacturer's quality control manual and facilities. Normally the review team consists of the representative from the Authorized Inspection Agency and the representative from the Jurisdiction. If, for whatever reason, the Jurisdiction does not participate in the review (as in the case of a foreign Manufacturer not located in a jurisdiction), then the review is conducted by the representative of the Authorized Inspection Agency and an individual chosen by the ASME known as the designee. If the Authorized Inspection Agency is also the Jurisdiction, then the review team consists of the representative of the jurisdiction and the ASME designee. The normal time for a review would be expected to be 1½ to 2 days, assuming thorough preparation and not many special circumstances. At the conclusion of the review, the team reports its findings and recommendations to the applicant's management before departing, and to the ASME via the ASME Qualification Review Report Form, submitted to the Subcommittee on Boiler and Pressure Vessel Accreditation of the ASME. In due course, the applicant is notified of the action taken by the committee on his application.

A Certificate of Authorization must be renewed every three years. It is the responsibility of the Manufacturer to request a renewal at least six months before the expiration date of his certificate in order to provide for a timely review process.

It is recommended that an applicant obtain the *National Board of Boiler and Pressure Vessel Inspectors Guide for ASME Code Sections I, IV and VIII, Divisions 1 and 2, Reviews and the National Board R Certificate of Authorization,* publication NB-57. This publication and others of the National Board may be obtained by writing to:

The National Board of Boiler and Pressure Vessel Inspectors
Order Department
1055 Crupper Avenue
Columbus, OH 43229
(614) 888-8320
FAX (614) 848-3474.

The *Guide* will be helpful in preparing for ASME and R stamp reviews.

As indicated above, the procedure for a safety valve manufacturer to apply for a stamp is somewhat different. There is no requirement that a safety valve manufacturer have an inspection agreement with an Authorized Inspection Agency. Therefore, the review team for the V stamp and certificate of authorization consists of an ASME designee. It should be noted that the valve manufacturer's quality control manual must include a provision for testing at a facility acceptable to the ASME (see A-302.14 and A-310 through A-316 of the Appendix to Section I). A list of approved safety valve test facilities is found at the end of this chapter.

The process of obtaining an R stamp and certificate of authorization is explained in Appendix C-R of the National Board Inspection Code (NBIC). The R stamp is required for organizations to perform repairs or alterations to boilers and pressure vessels. As with the ASME certificates, a contract with an Authorized Inspection Agency is required. (A list of Authorized Inspection Agencies offering inspection services for repairs and alterations is found at the end of this chapter.) The applicant must prepare a written quality control manual and implement the quality control system. The requirements for the quality control manual are given in 5.0 of Appendix C-R.

After the applicant (referred to in the NBIC as the organization) has prepared his quality control manual, he writes to the National Board to request a review. The review will be conducted by a team consisting of a representative of the Authorized Inspection Agency and a representative of the Qualifying Agency. The qualifying agency is the Jurisdiction in which the repair organization is located, unless:

(a) the organization is not located in a Jurisdiction.
(b) the Jurisdiction does not choose to do the review.
(c) the Jurisdiction is also the Authorized Inspection Agency.

In any of these three cases, the Qualifying Agency will be the National Board.

At the conclusion of the review, the team submits a report of its findings to the National Board, which then decides whether the R Certificate and stamp should be issued.

Upon writing to the National Board, an applicant will receive a letter explaining the fees and procedures for the Certificates of Authorization and the R stamps themselves, as well as the charges for the review team. He will also receive a copy of Appendix C-R and Appendix C-VR of the NBIC, and a copy of the publications catalog of the National Board. At the time of this writing, the charges for the R Certificate of Authorization are $510, and for the stamp itself, $60. The cost of the review itself is $100 per hour plus all expenses (hotel, meals, car rental, travel) of the reviewer or review team. A deposit of $2500 toward the cost of the review must be made in order to obtain a review date. The term of the Certificate is three years, and it is the responsibility of the repair organization to apply for renewal six months before the expiration date.

Appendix C-VR of the NBIC is also National Board Publication NB-65, the rules and procedures for obtaining the Certificate of Authorization to repair safety valves and safety relief valves. A contract with an Authorized Inspection Agency is not required. A written quality control manual is required, however, as is a survey of the facilities. The survey is performed by a representative of the National Board, with the jurisdiction being encouraged to participate as well. The decision as to whether to issue the Certificate of Authorization is made by the National Board Accreditation Committee on Repair of Pressure Relief Valves. Part of the review and evaluation process is the operational and capacity verification testing of repaired valves at an ASME-approved safety valve testing facility.

The adjustment of the set pressure of the safety valve and the setting of the blowdown are technically not repairs of the safety valve, and these adjustments may, with the authorization of the jurisdiction, be made by qualified employees of boiler users. This is stated in Appendix C-VR, paragraph 7.2; it is in apparent conflict with PG-72.3 of Section I of the ASME Code, which requires that a safety valve manufacturer or assembler perform the adjustment of the set pressure. The ASME requirement can be understood to apply to new construction only, with the National Board Inspection Code (and the Jurisdictions) governing the practices for safety valves on boilers and pressure vessels in service.

Enforcement official, address and telephone number

	Jurisdiction	Official	Address	City, state, zip code	Telephone no.
	Alabama	Mike Morgan	651 Folsom Administrative Bldg.	Montgomery, AL 36130	(205) 242-3460
	Alaska	Gerard F. Mankel	3301 Eagle Street, P.O. Box 107020	Anchorage, AK 99510-0720	(907) 269-4925
	Arizona	Thomas E. Rennie	P.O. Box 19070, 800 W. Washington Street	Phoenix, AZ 85007	(602) 542-1648
	Arkansas	Martin Skarda	1041 West Markham	Little Rock, AR 72205	(501) 682-4512
	California	John A. Lemire	1390 Market Street, Suite 910	San Francisco, CA 94102	(415) 557-1378
	Colorado	Donald Alexander	1120 Lincoln Street, Suite 1305	Denver, CO 80203-2140	(303) 894-7535
	Connecticut	James W. Corcoran	1111 Country Club Road	Middletown, CT 06457-9294	(203) 288-8034
	Delaware	Frank L. Gardner, Jr.	P.O. Box 674	Dover, DE 19903-0674	(302) 739-5889
7-95	Florida	Greg Lundberg	200 East Gaines Street	Tallahassee, FL 32399-0342	(904) 922-3171
	Georgia	Earl Everett	223 Courtland Street, NE, Suite 301	Atlanta, GA 30303	(404) 656-2966
	Idaho	Herb Carlson	Indus. Com. 317 Main Street, Statehouse Mail	Boise, ID 83720	(208) 334-6000
	Illinois	David A. Douin	1035 Stevenson Drive	Springfield, IL 62703-4259	(217) 785-1010
	Indiana	Anthony W. Meiring	402 W. Washington Street, Room C246	Indianapolis, IN 46204-2739	(317) 232-1921
	Iowa	Robert B. West	1000 East Grand Avenue	Des Moines, IA 50319-0209	(515) 281-3647
	Kansas	William E. Brown	5301 Goodman Lane	Overland Park, KS 66202-1110	(913) 296-4379
	Kentucky	Roger A. Coomes	1047 U.S. Route 127 South	Frankfort, KY 40601	(502) 564-3626
	Louisiana	Robert Cate	5150 Florida Boulevard	Baton Rouge, LA 70806	(504) 925-4911
7-95	Maine	VACANT	State Office Building, Station 45	Augusta, ME 04333-0045	(207) 624-6420
	Maryland	Myron H. Diehl, Jr.	501 St. Paul Place	Baltimore, MD 21202-2272	(301) 333-4160
	Massachusetts	John McCabe	One Ashburton Place, Room 1301	Boston, MA 02108	(617) 727-3200
	Michigan	Robert J. Aben, Jr.	7150 Harris Drive, P.O. Box 30254	Lansing, MI 48909	(517) 322-1836
7-95	Minnesota	Mark Rudak	443 Lafayette Road, 4th Floor	St. Paul, MN 55155-4304	(612) 296-1098

	Mississippi	Henry T. McEwen	2423 North State Street, Room 234	Jackson, MS 39215-1700	(601) 960-7918
7-95	Missouri	Darryl K. Peetz	1709 Industrial Drive	Jefferson City, MO 65109	(314) 751-8709
7-95	Montana	Timothy G. Gottsch	1218 East 6th Avenue, P.O. Box 200517	Helena, MT 59620-0517	(406) 444-6420
	Nebraska	David A. Starr	301 Centennial Mall South, Box 95024	Lincoln, NE 68509-5024	(402) 471-4721
	Nevada	Kenneth A. White	1370 S. Curry Street	Carson City, NV 89710	(702) 687-5249
	New Hampshire	Franklin F. Ellis	95 Pleasant Street, P.O. Box 2076	Concord, NH 03302-2076	(603) 271-2584
7-95	New Jersey	Raymond Snyder	CN 392	Trenton, NJ 08625-0392	(609) 292-2345
7-95	New Mexico	Vernon Cash	725 St. Michael's Drive, P.O. Box 25101	Santa Fe, NM 87505	(505) 827-7049
	New York	Ronald K. White	State Campus Building 12, Room 134	Albany, NY 12240-0102	(518) 457-2722
	North Carolina	Joseph W. Kapherr	4 West Edenton Street	Raleigh, NC 27601	(919) 733-2383
7-95	North Dakota	Robert Reetz	925 Basin Avenue	Bismarck, ND 58504-6647	(701) 328-9607
	Ohio	Dean T. Jagger	2323 W. Fifth Avenue, P.O. Box 825	Columbus, OH 43216	(614) 644-2236
7-95	Oklahoma	Thomas Monroe	4001 North Lincoln Boulevard	Oklahoma City, OK 73105-5212	(405) 528-1500
	Oregon	William C. Lundine	1535 Edgewater Drive, NW	Salem, OR 97310	(503) 373-7499
	Pennsylvania	John D. Payton	1520 Labor & Industry, 7th & Forster Streets	Harrisburg, PA 17120	(717) 787-2923
	Puerto Rico	Antonio Cárdenas	505 Muñoz Rivera Avenue	Hato Rey, PR 99018	(809) 754-2172
	Rhode Island	Thomas F. Wickham	610 Manton Avenue	Providence, RI 02909	(401) 457-1838
	South Carolina	William M. Lybrand	3600 Forest Drive, P.O. Box 11329	Columbia, SC 29211	(803) 734-9644
	South Dakota	Jerome D. Anderson	118 West Capitol Avenue	Pierre, SD 57501-2080	(605) 773-4311
	Tennessee	Danald E. Tanner	710 James Robertson Parkway	Nashville, TN 37243-0663	(615) 741-2123
	Texas	George Bynog	E. O. Thompson State Building, Box 12157	Austin, TX 78711	(512) 463-2904
	Utah	James C. Parsell	160 East 300 South, 3rd Floor	Salt Lake City, UT 84114-6620	(801) 530-6872
	Vermont	Malcolm J. Wheel	19 Commerce Street	Wiliston, VT 05495	(802) 658-2199

Enforcement official, address and telephone number (Continued)

Jurisdiction	Official	Address	City, state, zip code	Telephone no.
Virginia	Fred P. Barton	13 South Thirteenth Street	Richmond, VA 23219	(804) 786-3169
Washington	Richard Barkdoll	7273 Linderson Way SE, P.O. Box 44410	Olympia, WA 98504-4410	(206) 956-5270
West Virginia	Murry Shuff	106 Kinzer Street	Beckley, WV 25801	(304) 252-6995
Wisconsin	Virgil Kanable	201 E. Washington Avenue, Box 7969	Madison, WI 53707-7969	(608) 266-1904
Wyoming	Steve Foster	Herschler Building, 2 East	Cheyenne, WY 82002	(307) 777-7786
Canada				
Alberta	Dr. K. T. Lau	10808-99 Avenue, 6th Floor	Edmonton, AB T5K 0G5	(403) 427-6855
British Columbia	Lou Roussinos	#300-750 Pacific Boulevard	Vancouver, BC V6B 5E7	(604) 660-6243
Manitoba	I. Wayne Mault	500-401 York Avenue	Winnipeg, MB R3C 0P8	(204) 945-3374
New Brunswick	Dale E. Ross	470 York Street, P.O. Box 6000	Frederickton, NB E3B 3P7	(506) 453-4272
7-95 New Foundland & Labra.	E. Dennis Eastman	P.O. Box 8700	St. John's, NF A1B 4J6	(709) 729-2746
Northwest Territories	Lloyd Chase	4916 49 Street, P.O. Box 1320	Yellowknife, NT X1A 2L9	(403) 873-7399
Nova Scotia	Robert A. Yeo	5151 Terminal Road, P.O. Box 697	Halifax, NS B3J 2T8	(902) 424-7520
Ontario	H. D. Hanrath	3300 Bloor Street West, 3rd Floor W.	Toronto, ON M8X 2X4	(416) 234-6000
Prince Edward Island	W. A. Miller West	31 Gordon Drive, P.O. Box 2000	Charlottetown, PE C1A 7N8	(902) 368-5564
Québec	Mrs. Madiha Kotb	545 boul., Crémazie Est, 7e étage	Montréal, PQ H2M 2V2	(514) 873-6538
Saskatchewan	Nicholas Surtees	1855 Victoria Avenue	Regina, SK S4P 3V7	(306) 787-4509
Yukon Territory	Robert Speiss	P.O. Box 2703	Whitehorse, YT Y1A 2C6	(403) 667-5825
Cities and Counties				
Albuquerque	Michael R. Vanderslice	600 Second Street, NW, P.O. Box 1293	Albuquerque, NM 87103	(505) 764-1644

Buffalo	Tim Keenan	City Hall, Room 304	Buffalo, NY 14202	(716) 851-4909
Chicago	William J. Holmes	121 North LaSalle Street, Room 904	Chicago, IL 60602	(312) 744-7949
Denver	Domenic G. Stone	200 W. 14th Avenue	Denver, CO 80204	(303) 640-2031
Des Moines	Rick E. Bahanish	602 E. 1st Street	Des Moines, IA 50309	(515) 283-4946
Detroit	Athal (Al) Methvin	City-County Building, Room 408	Detroit, MI 48226	(313) 224-3205
Los Angeles	Jovie Aclaro	200 North Spring Street, Room 490	Los Angeles, CA 90012	(213) 485-3541
Miami	Mr. Gomez	275 NW, 2nd Street	Miami, FL 33128	(305) 350-7995
Milwaukee	Randal S. Pucek	841 N. Broadway, Room 1016	Milwaukee, WI 53202	(414) 278-2553
New Orleans	Glenn L. Ebeyer	1300 Perido Street, Room 7E04	New Orleans, LA 70112	(504) 565-6153
New York City	Dewey Hillman	60 Hudson Street	New York, NY 10013	(212) 312-8275
Omaha	W. D. Yatesil	Civic Center, 1819 Famam Street	Omaha, NE 68183	(402) 444-5383
St. Joseph	Mark Manville	City Hall, Room 201, 11th & Fredrick	St. Joseph, MO 64501	(816) 271-4855
St. Louis	D. J. Mayhew	1200 Market Street, Room 425	St. Louis, MO 63103	(314) 622-3375
Seattle	Donald W. Gentry	710 2nd Avenue, Suite 700	Seattle, WA 98104-1703	(206) 684-8459
Spokane	Don Hammond	W. 808 Spokane Falls Boulevard	Spokane, WA 99201-3325	(509) 625-6133
Tacoma	Richard L. Boney	747 Market Street, Suite 345	Tacoma, WA 98402	(206) 591-5029
Tucson	Ron Lohrman	201 N. Stone Avenue, P.O. Box 27210	Tucson, AZ 85726-7210	(602) 791-5550
University City	Harold Burkemper	6801 Delmar Boulevard	University City, MO 63130	(314) 862-0770
White Plains	Peter Mulvey	255 Main Street	White Plains, NY 10601	(914) 422-1306
Dade County	John Davidson	111 NW 1st Street, Suite 1310	Miami, FL 33128-1974	(305) 375-2787
Jefferson Parish	Joseph Calcagno	1221 Elmwood Park Boulevard	Harahan, LA 70123	(504) 736-6918
St. Louis County	David H. Pollmann	41 South Central	Clayton, MO 63105	(314) 889-2862
District of Columbia	Alex Colvin, II	614 H. Street, NW, Room LL9	Washington, DC 20001	(202) 727-7506

List of authorized inspection agencies for new construction.

ABS Industrial Verification, Inc.
16855 Northchase Drive
Houston, TX 77060
(713) 873-5200
FAX (713) 874-9553

American States Insurance Co.
Boiler & Machinery Department
P.O. Box 1636
Indianapolis, IN 46206
(317) 262-6286
FAX (317) 262-6391

Arise, Inc.
Corporate Plaza
6480 Rockside Woods Blvd. South
Cleveland, OH 44131-2206
(216) 447-1600
FAX (216) 642-4381

Boiler Inspection and Insurance
Company of Canada
18 King Street East
Toronto, ON Canada M5C 1C4
(416) 363-5491
FAX (416) 363-0538

Chubb Group of Insurance Companies
Inspection Services
Warren, NJ 07061
(908) 580-3695
FAX (908) 580-3543

Hartford Steam Boiler Inspection
& Insurance Company
One State Street
Hartford, CT 06102
(203) 722-5662
FAX (203) 722-5530

Industrial Risk Insurers
National B/M Loss Prevention
85 Woodland Street
Hartford, CT 06102
(203) 520-6037
FAX (203) 549-5780

Commercial Union Insurance Co.
Contract Inspection Services
One Beacon Street
Boston, MA 02108-3100
(617) 725-7309
FAX (617) 725-6094

Delta Lloyds Insurance Co.
Inspection Services Division
P.O. Box 165
Houston, TX 77001-0165
(713) 666-5556
FAX (713) 666-4006

Factory Mutual Engineering Association
1151 Boston-Providence Tpke.
P.O. Box 9102
Norwood, MA 02062
(617) 255-4272
FAX (617) 762-9375

Gerling Global General
Global House
Boiler & Machinery Department
480 University Avenue
Toronto, ON Canada M5G 1V6
(416) 598-4651
FAX (416) 598-9507

Cincinnati Insurance Company
P.O. Box 145496
Cincinnati, OH 45250-9988
(513) 870-2681
FAX (513) 870-2059

Indiana Insurance
Boiler & Machinery Department
320 E. 96th Street
P.O. Box 1967
Indianapolis, IN 46206-1967
(317) 581-6400
FAX (317) 581-6451

Kemper National Insurance Companies
Loss Control Engineering Dept. D-3
One Kemper Drive
Long Grove, IL 60049
(708) 320-3102
FAX (708) 320-4417

LR Insurance, Inc.
17 Battery Place
New York, NY 10004
(212) 425-8050
FAX (212) 363-9610

Old Republic Insurance Co.
690 E. Lamar Blvd., Ste 580
Arlington, TX 76011
(817) 543-8032
FAX (817) 543-8036

Reliance Insurance Companies
Boiler & Machinery/Loss Control
77 Water Street
New York, NY 10005
(212) 858-6323
FAX (212) 858-6550

Royal Insurance of Canada
Engineering & Technical Services
10 Wellington Street East
Toronto, ON Canada M5E 1L5
(416) 366-7511
FAX (416) 367-9869

Western National Mutual
Insurance Company
5350 West 78th Street
P.O. Box 1463
Minneapolis, MN 55440
(612) 835-5350
FAX (612) 921-3163

Liberty Mutual Insurance Company
Loss Control Prevention Department
Boiler & Machinery Section
175 Berkeley Street
Boston, MA 02117-0140
(617) 357-9500
FAX (617) 695-9216

Ping An Insurance (USA) Co. Ltd.
Beijing Ping An Technical
Inspection Co.
564 to 567 Jing Tong Hotel
No. B27, Wan Shou Lu Street
Beijing 100036, China
Phone 6841029
FAX 8233453

Royal Insurance
Loss Control
9300 Arrowpoint Blvd.
P.O. Box 1000
Charlotte, NC 28210-1000
(704) 522-2932
FAX (704) 522-3200

The Travelers
Boiler & Machinery Engineering
One Tower Square
Hartford, CT 06183-2032
(203) 277-3081
FAX (203) 954-7958

Zurich American Insurance Group
Tower 1-10th Floor
1400 American Lane
Schaumberg, IL 60196
(708) 605-7643
FAX (708) 605-7934

The American Society of Mechanical Engineers

CERTIFICATE OF AUTHORIZATION

This certificate accredits the named company as authorized to use the indicated symbol of the American Society of Mechanical Engineers (ASME) for the scope of activity shown below in accordance with the applicable rules of the ASME Boiler and Pressure Vessel Code. The use of the code symbol and the authority granted by this Certificate of Authorization are subject to the provisions of the agreement set forth in the application. Any construction stamped with this symbol shall have been built strictly in accordance with the provisions of the ASME Boiler and Pressure Vessel Code.

COMPANY

HELFRICH BROS. BOILER WORKS, INC.
39 MERRIMACK STREET
LAWRENCE, MASSACHUSETTS 01843

SCOPE

POWER BOILERS AT THE ABOVE LOCATION
AND FIELD SITES CONTROLLED BY THIS LOCATION

AUTHORIZED	FEBRUARY 20, 1995
EXPIRES	MARCH 8, 1998
CERTIFICATE NUMBER	12,147
SYMBOL	S

CHAIRMAN OF THE BOILER
AND PRESSURE VESSEL COMMITTEE

DIRECTOR, ACCREDITATION AND CERTIFICATION

ASME ®

Certificate of authorization for the S stamp.

Accepted safety valve testing laboratories.

Anderson, Greenwood & Company
(mailing address)
3950 Greenbriar
Stafford, TX 77477
(713) 274-4400
FAX (713) 240-1800

Continental Disk Corporation
3160 West Heartland Drive
Liberty, MO 64068
(816) 792-1500
FAX (816) 792-2277

Crosby Valve & Gage Company
43 Kendrick Street
Wrentham, MA 02093
(508) 384-3121
FAX (508) 384-8675

Dresser Industries, Inc.
P.O. Box 1430
3225 North Highway
Alexandria, LA 71301
(318) 640-6082
FAX (318) 640-6175

Engineered Controls International, Inc.
100 Rego Drive
Elon College, NC 27244
(919) 449-7707
FAX (919) 449-6594

Kunkle Industries, Inc.
8222 Bluffton Road
Fort Wayne IN 46809
(mailing address) P.O. Box 1740
Fort Wayne, IN 46801
(219) 747-3405
FAX (219) 747-7958

Leser GMBH & Co. KG
Wendenstrasse 133
D-20537 Hamburg, Germany
(49) 40-25-15-00-50
FAX (49) 40-25-15-00-24

The National Board of Boiler & Pressure
Vessel Inspectors (mailing address)
1055 Crupper Avenue
Columbus, OH 43229
(614) 888-8320
FAX (614) 848-3474

Teledyne Faris Engineering
400 Commercial Avenue
Palisades Park, NJ 07650
(201) 944-6300
FAX (201) 944-4315

Photo 10.1 R-stamp nameplate on a repaired boiler. Photo by the authors.

Authorized inspection agencies for inservice inspection.

ABS Industrial Verification Inc.
16855 Northchase Drive
Houston, Texas 77060
(713) 873-5200
FAX (713) 874-9553

American States Insurance Company
Boiler and Machinery Department
P.O. Box 1636
Indianapolis, IN 46206
(317) 262-6286
FAX (317) 262-6391

Arise Incorporated
Corporate Plaza II, Suite 110
6480 Rockside Woods Blvd.
Cleveland, OH 43131-2206
(216) 447-1600
FAX (216) 642-4381

Boiler Inspection and Insurance Company of Canada
18 King Street East (2nd Floor)
Toronto, ON Canada M5C 1C4
(416) 363-5491
FAX (416) 363-0538

Boreal Generale
1100 West Rene-Levesque Blue
Montreal, PQ Canada N3B 4P4
(514) 392-6000
FAX (514) 392-6044

Chubb Group of Insurance Companies
Inspection Services
P.O. Box 1615
Warren, NJ 07061
(908) 580-3695
FAX (908) 580-3543

Factory Mutual Engineering Association
1151 Boston Providence Turnpike
P.O. Box 9102
Norwood, MA 02062
(617) 762-4272
FAX (617) 762-9375

Cigna
Property & Casualty
1601 Chestnut St. TLT-35
P.O. Box 7716
Philadelphia, PA 19192-2357
(215) 761-1727
FAX (215) 761-2205

Cincinnati Insurance Company
P.O. Box 145496
Cincinnati, OH 43250-9988
(513) 870-2681
FAX (513) 870-2059

Commercial Union Assurance Co. of Canada
Commercial Union Tower,
Toronto Dominion Center
P.O. Box 441
Toronto, ON Canada M5K 1L9
(416) 361-2500
FAX (416) 361-2606

Commercial Union Assurance Co.
Contract Inspection Services
One Beacon Street
Boston, MA 02108-3100
(617) 725-7309
FAX (617) 725-6094

Continental Insurance Company
C-Tek Technical Services
Boiler and Machinery Department
One Continental Drive
Cranbury, NJ 08570-2105
(609) 395-2357
FAX (609) 395-5898

Delta Lloyds Insurance Company
Inspection Services Division
P.O. Box 165
Houston, TX 77001-0165
(713) 666-5556
FAX (713) 666-4006

Kemper National Insurance Companies
Loss Control Engineering Dept., D-3
One Kemper Drive
Long Grove, IL 60049
(708) 320-3101
FAX (708) 320-4417

Gerling Global General
Global House
Boiler & Machinery Department
480 University Ave.
Toronto, ON Canada M5G 1V6
(416) 598-4651
FAX (416) 598-9507

Hartford Steam Boiler Inspection & Insurance Company
One State Street
Hartford, CT 06102
(203) 722-5662
FAX (203) 722-5530

Indiana Insurance
Boiler & Machinery Department
320 E. 96th Street
P.O. Box 1967
Indianapolis, IN 46206-1967
(317) 581-6400
FAX (317) 581-6451

Industrial Risk Insurers
Attn: National B/M Loss Prevention
85 Woodland Street
Hartford, CT 06102
(203) 520-6037
FAX (203) 549-5780

Royal Insurance
Loss Control
9300 Arrowpoint Boulevard
P.O. Box 1000
Charlotte, NC 28210-1000
(704) 522-2932
FAX (704) 522-3200

The Travelers
Boiler and Machinery Division
One Tower Square (7SB)
Hartford, CT 06183-2032
(203) 277-3081
FAX (203) 954-7958

Western National Mutual Insurance Company
5350 West 78th Street
P.O. Box 1463
Minneapolis, MN 55440
(612) 835-5350
FAX (612) 921-3163

LR Insurance, Inc.
17 Battery Place
New York, NY 10004
(212) 425-8050
FAX (212) 363-9610

Old Republic Insurance Company
690 E. Lamar Blvd., Ste 580
Arlington, TX 76011
(817) 543-8023
FAX (817) 543-8036

Ping An Insurance (U.S.A.) Co. Ltd.
Beijing Ping An Technical Inspection Co.
564 to 567 Jing Tong Hotel
No. B27, Wan Shou Lu Street
Bejiing 100036, China
Phone 6841029
FAX 8233453

Reliance Insurance Companies
Boiler & Machinery/Loss Control
77 Water Street
New York, NY 10005
(212) 858-6323
FAX (212) 858-6550

Royal Insurance of Canada
Engineering & Technical Services
10 Wellington Street East
Toronto, ON Canada M5E 1L5
(416) 366-7511
FAX (416) 367-9869

Wellington Insurance Company
Loss Control/Boiler & Machinery
Suite 2102
20 Queen Street West
Toronto, ON Canada M5H 3R3
(416) 581-1683
FAX (416) 581-1759

Zurich American Insurance Group
Tower 1-10th Floor
1400 American Lane
Schaumberg, IL 60196
(708) 605-7643
FAX (708) 605-7934

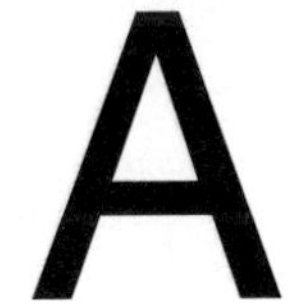

Appendix A

Table of Pipe Sizes

All material contained in this publication is for general information only. This material should not, therefore, be used or relied upon for any specific application without independent competent professional examination and verification of its accuracy, suitability and applicability. Anyone making use of the material does so at his own risk and assumes any and all liability resulting from such use. U.S. Steel Group of USX Corporation disclaims any and all express or implied warranties of fitness for any general or particular purpose.

USS Standard Pipe* and Line Pipe† Data

Reprinted by permission of USS Tubular Products.

Size		Nom. Wall Thickness, Inches	Weight Lb/Ft Plain End	Weight Tons/ Mile	Class	Sched. No.	Mill Hydrostatic Test Pressure‡ — psi						
NPS	OD Inches						Grade A	Grade B	Grade X-42	Grade X-46	Grade X-52	Grade X-56	Grade X-60
1½	1.900 (Seamless)	0.145	2.718		STD			1300					
		0.200	3.631		XS			1900					
		0.281	4.839					2050					
2	2.375 (Seamless)	0.154	3.65	9.64	STD	40	2330	2500	3000	3000	3000	3000	3000
		0.218	5.02	13.26	XS	80	2500	2500	3000	3000	3000	3000	3000
		0.344	7.46	19.70		160	2500	2500	3000	3000	3000	3000	3000
		0.436	9.03	23.84	XXS		2500	2500	3000	3000	3000	3000	3000
2½	2.875 (Seamless)	0.203	5.79	15.29	STD	40	2500	2500	3000	3000	3000	3000	3000
		0.276	7.66	20.23	XS	80	2500	2500	3000	3000	3000	3000	3000
		0.375	10.01	26.43		160	2500	2500	3000	3000	3000	3000	3000
		0.552	13.69	36.15	XXS		2500	2500	3000	3000	3000	3000	3000
3	3.500 (Seamless)	0.216	7.58	20.00	STD	40	2220	2500	3000	3000	3000	3000	3300
		0.250	8.68	22.91			2500	2500	3000	3000	3000	3000	3000
		0.281	9.66	25.50			2500	2500	3000	3000	3000	3000	3000
		0.300	10.25	27.07	XS	80	2500	2500	3000	3000	3000	3000	3000
		0.438	14.32	37.81		160	2500	2500	3000	3000	3000	3000	3000
		0.600	18.58	49.06	XXS		2500	2500	3000	3000	3000	3000	3000
3½	4.000 (Seamless)	0.226	9.11	24.05	STD	40	2030	2370	3000	3000	3000	3000	3000
		0.250	10.01	26.43			2250	2620	3000	3000	3000	3000	3000
		0.281	11.16	29.47			2530	2800	3000	3000	3000	3000	3000
		0.318	12.50	33.01	XS	80	2800	2800	3000	3000	3000	3000	3000
		0.636	22.85	60.32			2800	2800	3000	3000	3000	3000	3000
4	4.500 (Seamless and Electric Resistance Weld)	0.156	7.24	19.11			1250	1460	2180	2390	2700	2910	3000
		0.172	7.95	20.99			1380	1610	2410	2640	2980	3000	3000
		0.188	8.66	22.86			1500	1750	2630	2880	3000	3000	3000
		0.203	9.32	24.59			1620	1890	2840	3000	3000	3000	3000
		0.219	10.01	26.43			1750	2040	3000	3000	3000	3000	3000
		0.237	10.79	28.49	STD	40	1900	2210	3000	3000	3000	3000	3000
		0.250	11.35	29.96			2000	2330	3000	3000	3000	3000	3000
		0.281	12.66	33.43			2250	2620	3000	3000	3000	3000	3000
		0.312	13.96	36.84			2500	2800	3000	3000	3000	3000	3000
		0.337	14.98	39.56	XS	80	2700	2800	3000	3000	3000	3000	3000
		0.438	19.00	50.16		120	2800	2800	3000	3000	3000	3000	3000
		0.531	22.51	59.42		160	2800	2800	3000	3000	3000	3000	3000
		0.674	27.54	72.71	XXS		2800	2800	3000	3000	3000	3000	3000

Note: Some outside diameters, walls and grades are listed for information only and are not necessarily regular production items. Sizes not listed may be negotiated upon inquiry.

* Standard pipe is manufactured to comply with ASTM standards, such as A53 and A106.

† Line pipe is manufactured to meet API standard 5L.

‡ Mill Hydrostatic Test Pressure Data. The mill hydrostatic test pressures indicated were calculated on the basis of a fiber stress equal to a percentage of the specified minimum yield strength for the various sizes and grades. For specific information, ASTM Standards and/or API Specification 5L should be consulted. Due to limited pump capacity, maximum hydrostatic test pressures for 22-inch, 24-inch, and 26-inch OD seamless are 2700, 2300, and 2000 psi, respectively.

USS Standard Pipe* and Line Pipe† Data (*Continued*)

Size		Nom. Wall Thick-ness, Inches	Weight Lb/Ft Plain End	Weight Tons/ Mile	Class	Sched. No.	Mill Hydrostatic Test Pressure‡ — psi						
NPS	OD Inches						Grade A	Grade B	Grade X-42	Grade X-46	Grade X-52	Grade X-56	Grade X-60
5	5.563 (Seamless and Electric Resistance Weld)	0.188	10.79	28.49			1220	1420	2130	2330	2640	2840	3000
		0.219	12.50	33.00			1420	1650	2480	2720	3000	3000	3000
		0.258	14.62	38.59	STD	40	1670	1950	2920	3000	3000	3000	3000
		0.281	15.85	41.85			1820	2120	3000	3000	3000	3000	3000
		0.312	17.50	46.19			2020	2360	3000	3000	3000	3000	3000
		0.344	19.17	50.62			2230	2600	3000	3000	3000	3000	3000
		0.375	20.78	54.85	XS	80	2430	2800	3000	3000	3000	3000	3000
		0.500	27.04	71.38		120	2800	2800	3000	3000	3000	3000	3000
		0.625	32.96	87.02		160	2800	2800	3000	3000	3000	3000	3000
		0.750	38.55	101.78	XXS		2800	2800	3000	3000	3000	3000	3000
		0.875	43.81	115.66			2800	2800	3000	3000	3000	3000	3000
		0.938	46.33	122.32			2800	2800	3000	3000	3000	3000	3000
6	6.625 (Seamless and Electric Resistance Weld)	0.156	10.78	28.45			1060	1240	1480	1620	1840	1980	2120
		0.172	11.85	31.29			1170	1360	1640	1790	2030	2180	2340
		0.188	12.92	34.12			1280	1490	1790	1960	2210	2380	2550
		0.203	13.92	36.76			1380	1610	1930	2110	2390	2570	2760
		0.219	14.98	39.56			1490	1740	2080	2280	2580	2780	2980
		0.250	17.02	44.94			1700	1980	2380	2600	2940	3000	3000
		0.280	18.97	50.09	STD	40	1900	2220	2660	2920	3000	3000	3000
		0.312	21.04	55.53			2120	2470	2970	3000	3000	3000	3000
		0.344	23.08	60.92			2340	2730	3000	3000	3000	3000	3000
		0.375	25.03	66.08			2550	2800	3000	3000	3000	3000	3000
		0.432	28.57	75.43	XS	80	2800	2800	3000	3000	3000	3000	3000
		0.500	32.71	86.35			2800	2800	3000	3000	3000	3000	3000
		0.562	36.39	96.07		120	2800	2800	3000	3000	3000	3000	3000
		0.625	40.05	105.73			2800	2800	3000	3000	3000	3000	3000
		0.719	45.35	119.73		160	2800	2800	3000	3000	3000	3000	3000
		0.864	53.16	140.34	XXS		2800	2800	3000	3000	3000	3000	3000
		1.000	60.08	158.60			2800	2800	3000	3000	3000	3000	3000
		1.125	66.08	174.46			2800	2800	3000	3000	3000	3000	3000
		1.188	68.98	182.11			2800	2800	3000	3000	3000	3000	3000

USS Standard Pipe* and Line Pipe† Data (*Continued*)

Size		Nom. Wall Thickness, Inches	Weight Lb/Ft Plain End	Weight Tons/Mile	Class	Sched. No.	Mill Hydrostatic Test Pressure‡ — psi								
NPS	OD Inches						Grade A	Grade B	Grade X-42	Grade X-46	Grade X-52	Grade X-56	Grade X-60	Grade X-65	Grade X-70
8	8.625 (Seamless and Electrical Resistance Weld)	0.188	16.94	44.72			980	1140	1370	1500	1700	1830	1960	2130	2290
		0.203	18.26	48.20			1060	1240	1480	1620	1840	1980	2120	2290	2470
		0.219	19.66	51.90			1140	1330	1600	1750	1980	2130	2290	2480	2670
		0.250	22.36	59.03		20	1300	1520	1830	2000	2260	2430	2610	2830	3000
		0.277	24.70	65.20		30	1450	1690	2020	2220	2510	2700	2890	3000	3000
		0.312	27.70	73.13			1630	1900	2280	2500	2820	3000	3000	3000	3000
		0.322	28.55	75.38	STD	40	1680	1960	2350	2580	2910	3000	3000	3000	3000
		0.344	30.42	80.32			1790	2090	2510	2750	3000	3000	3000	3000	3000
		0.375	33.04	87.23			1960	2280	2740	3000	3000	3000	3000	3000	3000
		0.406	35.64	94.08		60	2120	2470	2970	3000	3000	3000	3000	3000	3000
		0.438	38.30	101.11			2290	2670	3000	3000	3000	3000	3000	3000	3000
		0.500	43.39	114.54	XS	80	2610	2800	3000	3000	3000	3000	3000	3000	3000
		0.562	48.40	127.76			2800	2800	3000	3000	3000	3000	3000	3000	3000
		0.594	50.95	134.50		100	2800	2800	3000	3000	3000	3000	3000	3000	3000
		0.625	53.40	140.98			2800	2800	3000	3000	3000	3000	3000	3000	3000
		0.719	60.71	160.27		120	2800	2900	3000	3000	3000	3000	3000	3000	3000
		0.812	67.76	178.87		140	2800	2800	3000	3000	3000	3000	3000	3000	3000
		0.875	72.42	191.20	XXS		2800	2800	3000	3000	3000	3000	3000	3000	3000
		0.906	74.69	197.18		160	2800	2800	3000	3000	3000	3000	3000	3000	3000
		1.000	81.44	214.99			2800	2800	3000	3000	3000	3000	3000	3000	3000
		1.125	90.11	237.90			2800	2800	3000	3000	3000	3000	3000	3000	3000
		1.250	98.46	259.93			2800	2800	3000	3000	3000	3000	3000	3000	3000
10	10.750 (Seamless and Electric Resistance Weld)	0.188	21.21	55.99			790	920	1250	1370	1550	1660	1780	1930	2080
		0.203	22.87	60.37			850	990	1350	1480	1670	1800	1930	2090	2250
		0.219	24.63	65.03			920	1070	1450	1590	1800	1940	2080	2250	2420
		0.250	28.04	74.01		20	1050	1220	1660	1820	2060	2210	2370	2570	2770
		0.279	31.20	82.37			1170	1360	1850	2030	2290	2470	2650	2870	3000
		0.307	34.24	90.39		30	1290	1500	2040	2230	2520	2720	2910	3000	3000
		0.344	38.23	100.93			1440	1680	2280	2500	2830	3000	3000	3000	3000
		0.365	40.48	106.87	STD	40	1530	1780	2420	2660	3000	3000	3000	3000	3000
		0.438	48.24	127.35			1830	2140	2910	3000	3000	3000	3000	3000	3000
		0.500	54.74	144.50	XS	60	2090	2440	3000	3000	3000	3000	3000	3000	3000
		0.562	61.15	161.44			2350	2740	3000	3000	3000	3000	3000	3000	3000
		0.594	64.43	170.09		80	2490	2800	3000	3000	3000	3000	3000	3000	3000
		0.625	67.58	178.42			2620	2800	3000	3000	3000	3000	3000	3000	3000
		0.719	77.03	203.35		100	2800	2800	3000	3000	3000	3000	3000	3000	3000
		0.812	86.18	227.53			2800	2800	3000	3000	3000	3000	3000	3000	3000
		0.844	89.29	235.73		120	2800	2800	3000	3000	3000	3000	3000	3000	3000
		1.000	104.13	274.90	XXS	140	2800	2800	3000	3000	3000	3000	3000	3000	3000
		1.125	115.64	305.30		160	2800	2800	3000	3000	3000	3000	3000	3000	3000
		1.438	143.01	377.55			2800	2800	3000	3000	3000	3000	3000	3000	3000

USS Standard Pipe* and Line Pipe† Data (*Continued*)

Size		Nom. Wall Thick-ness, Inches	Weight Lb/Ft Plain End	Weight Tons/ Mile	Class	Sched. No.	Mill Hydrostatic Test Pressure‡ — psi								
NPS	OD Inches						Grade A	Grade B	Grade X-42	Grade X-46	Grade X-52	Grade X-56	Grade X-60	Grade X-65	Grade X-70
12	12.750 (Seamless and Electric Resistance Weld)	0.188	25.22	66.59			660	770	1050	1150	1300	1400	1500	1630	1750
		0.203	27.20	71.81			720	840	1140	1250	1410	1520	1620	1760	1890
		0.219	29.31	77.38			770	900	1230	1340	1520	1640	1750	1900	2040
		0.250	33.38	88.11		20	880	1030	1400	1530	1730	1870	2000	2170	2330
		0.281	37.42	98.79			990	1160	1570	1720	1950	2100	2250	2440	2620
		0.312	41.45	109.42			1100	1280	1750	1910	2160	2330	2500	2700	2910
		0.330	43.77	115.56		30	1160	1360	1850	2020	2290	2460	2640	2860	3000
		0.344	45.58	120.33			1210	1420	1930	2110	2390	2570	2750	2980	3000
		0.375	49.56	130.84	STD		1320	1540	2100	2300	2600	2800	3000	3000	3000
		0.406	53.52	141.30		40	1430	1670	2270	2490	2810	3000	3000	3000	3000
		0.438	57.59	152.05			1550	1800	2450	2690	3000	3000	3000	3000	3000
		0.500	65.42	172.70	XS		1760	2060	2800	3000	3000	3000	3000	3000	3000
		0.562	73.15	193.13		60	1980	2310	3000	3000	3000	3000	3000	3000	3000
		0.625	80.93	213.67			2210	2570	3000	3000	3000	3000	3000	3000	3000
		0.688	88.63	233.98		80	2430	2800	3000	3000	3000	3000	3000	3000	3000
		0.750	96.12	253.76			2650	2800	3000	3000	3000	3000	3000	3000	3000
		0.844	107.32	283.32		100	2800	2800	3000	3000	3000	3000	3000	3000	3000
		1.000	125.49	331.29	XXS	120	2800	2800	3000	3000	3000	3000	3000	3000	3000
		1.125	139.67	368.74		140	2800	2800	3000	3000	3000	3000	3000	3000	3000
		1.312	160.27	423.12		160	2800	2800	3000	3000	3000	3000	3000	3000	3000
		1.500	180.22	475.79			2800	2800	3000	3000	3000	3000	3000	3000	3000
		1.594	189.92	501.39			2800	2800	3000	3000	3000	3000	3000	3000	3000
		1.625	193.07	509.72			2800	2800	3000	3000	3000	3000	3000	3000	3000
		2.000	229.62	606.20			2800	2800	3000	3000	3000	3000	3000	3000	3000
		2.250	252.32	666.12			2800	2800	3000	3000	3000	3000	3000	3000	3000

USS Standard Pipe* and Line Pipe† Data (*Continued*)

Size		Nom. Wall Thick-ness, Inches	Weight Lb/Ft Plain End	Weight Tons/Mile	Class	Sched. No.	Mill Hydrostatic Test Pressure‡ — psi								
NPS	OD Inches						Grade A	Grade B	Grade X-42	Grade X-46	Grade X-52	Grade X-56	Grade X-60	Grade X-65	Grade X-70
14	14.000 (Seamless and Electric Resistance Weld)	0.219	32.23	85.09			700	820	1120	1220	1380	1490	1600	1730	1860
		0.250	36.71	96.92			800	940	1280	1400	1580	1700	1820	1970	2130
		0.281	41.17	108.69			900	1050	1430	1570	1770	1910	2050	2220	2390
		0.312	45.61	120.41		20	1000	1170	1590	1740	1970	2120	2270	2460	2650
		0.344	50.17	132.45			1110	1290	1750	1920	2170	2340	2510	2720	2920
		0.375	54.57	144.06	STD	30	1210	1410	1910	2090	2370	2550	2730	2960	3000
		0.406	58.94	155.61			1300	1520	2070	2270	2560	2760	2960	3000	3000
		0.438	63.44	167.48		40	1410	1640	2230	2450	2770	2980	3000	3000	3000
		0.469	67.78	178.93			1510	1760	2390	2620	2960	3000	3000	3000	3000
		0.500	72.09	190.32	XS		1610	1880	2550	2790	3000	3000	3000	3000	3000
		0.562	80.66	212.93			1810	2110	2870	3000	3000	3000	3000	3000	3000
		0.594	85.05	224.52		60	1910	2230	3000	3000	3000	3000	3000	3000	3000
		0.625	89.28	235.69			2010	2340	3000	3000	3000	3000	3000	3000	3000
		0.688	97.81	258.23			2210	2580	3000	3000	3000	3000	3000	3000	3000
		0.750	106.13	280.19		80	2410	2800	3000	3000	3000	3000	3000	3000	3000
		0.812	114.37	301.93			2610	2800	3000	3000	3000	3000	3000	3000	3000
		0.938	130.85	345.45		100	2800	2800	3000	3000	3000	3000	3000	3000	3000
		1.094	150.79	398.09		120	2800	2800	3000	3000	3000	3000	3000	3000	3000
		1.250	170.21	449.36		140	2800	2800	3000	3000	3000	3000	3000	3000	3000
		1.406	189.11	499.26		160	2800	2800	3000	3000	3000	3000	3000	3000	3000
		1.656	218.32	576.36			2800	2800	3000	3000	3000	3000	3000	3000	3000
		1.750	228.95	604.43			2800	2800	3000	3000	3000	3000	3000	3000	3000
		2.000	256.32	676.68			2800	2800	3000	3000	3000	3000	3000	3000	3000
16	16.000 (Seamless and Electric Resistance Weld)	0.219	36.91	97.44			620	720	980	1070	1210	1300	1400	1510	1630
		0.250	42.05	111.02			700	820	1120	1220	1380	1490	1590	1730	1860
		0.281	47.17	124.54			790	920	1250	1370	1550	1670	1790	1940	2090
		0.312	52.27	138.01		20	880	1020	1390	1520	1720	1860	1990	2150	2320
		0.344	57.52	151.85			970	1130	1540	1680	1900	2050	2190	2380	2560
		0.375	62.58	165.21	STD	30	1050	1230	1670	1830	2070	2230	2390	2590	2790
		0.406	67.62	178.51			1140	1330	1810	1980	2240	2420	2590	2800	3000
		0.438	72.80	192.18			1230	1440	1950	2140	2420	2610	2790	3000	3000
		0.469	77.79	205.37			1320	1540	2090	2290	2590	2790	2990	3000	3000
		0.500	82.77	218.51	XS	40	1410	1640	2230	2440	2760	2980	3000	3000	3000
		0.562	92.66	244.63			1580	1840	2510	2750	3000	3000	3000	3000	3000
		0.625	102.63	270.94			1760	2050	2790	3000	3000	3000	3000	3000	3000
		0.656	107.50	283.80		60	1840	2150	2930	3000	3000	3000	3000	3000	3000
		0.688	112.51	297.03			1940	2260	3000	3000	3000	3000	3000	3000	3000
		0.750	122.15	322.48			2110	2460	3000	3000	3000	3000	3000	3000	3000
		0.812	131.71	347.72			2280	2660	3000	3000	3000	3000	3000	3000	3000
		0.844	136.61	360.66		80	2370	2770	3000	3000	3000	3000	3000	3000	3000
		1.031	164.82	435.14		100	2800	2800	3000	3000	3000	3000	3000	3000	3000
		1.219	192.43	508.02		120	2800	2800	3000	3000	3000	3000	3000	3000	3000
		1.438	223.64	590.41		140	2800	2800	3000	3000	3000	3000	3000	3000	3000
		1.594	245.25	647.45		160	2800	2800	3000	3000	3000	3000	3000	3000	3000
		1.618	248.52	656.09			2800	2800	3000	3000	3000	3000	3000	3000	3000
		1.875	282.85	746.72			2800	2800	3000	3000	3000	3000	3000	3000	3000
		2.000	299.04	789.47			2800	2800	3000	3000	3000	3000	3000	3000	3000

USS Standard Pipe* and Line Pipe† Data (*Continued*)

Size		Nom. Wall Thickness, Inches	Weight Lb/Ft Plain End	Weight Tons/ Mile	Class	Sched. No.	Mill Hydrostatic Test Pressure‡ — psi								
NPS	OD Inches						Grade A	Grade B	Grade X-42	Grade X-46	Grade X-52	Grade X-56	Grade X-60	Grade X-65	Grade X-70
18	18.000 (Seamless and Electric Resistance Weld)	0.250	47.39	125.12			620	730	990	1090	1230	1320	1420	1530	1650
		0.281	53.18	140.39			700	820	1110	1220	1380	1490	1590	1730	1860
		0.312	58.94	155.60		20	780	910	1240	1360	1530	1650	1770	1920	2060
		0.344	64.87	171.25			860	1000	1360	1490	1690	1820	1950	2110	2270
		0.375	70.59	186.35	STD		940	1090	1490	1630	1840	1980	2120	2300	2480
		0.406	76.29	201.40			1010	1180	1610	1760	1990	2150	2300	2490	2680
		0.438	82.15	216.88		30	1100	1280	1740	1900	2150	2320	2480	2690	2900
		0.469	87.81	231.82			1170	1370	1860	2040	2300	2480	2660	2880	3000
		0.500	93.45	246.71	XS		1250	1460	1980	2170	2460	2640	2830	3000	3000
		0.562	104.67	276.32		40	1400	1640	2230	2440	2760	2970	3000	3000	3000
		0.625	115.98	306.18			1560	1820	2480	2720	3000	3000	3000	3000	3000
		0.688	127.21	335.82			1720	2010	2730	2990	3000	3000	3000	3000	3000
		0.750	138.17	364.78		60	1880	2190	2980	3000	3000	3000	3000	3000	3000
		0.812	149.06	393.51			2030	2370	3000	3000	3000	3000	3000	3000	3000
		0.938	170.92	451.24		80	2340	2740	3000	3000	3000	3000	3000	3000	3000
		1.156	207.96	549.01		100	2800	2800	3000	3000	3000	3000	3000	3000	3000
		1.375	244.14	644.52		120	2800	2800	3000	3000	3000	3000	3000	3000	3000
		1.562	274.22	723.94		140	2800	2800	3000	3000	3000	3000	3000	3000	3000
20	20.000 (Seamless and Electric Resistance Weld)	0.250	52.73	139.21			560	660	940	1040	1170	1260	1350	1460	1580
		0.281	59.18	156.23			630	740	1060	1160	1320	1420	1520	1640	1770
		0.312	65.60	173.19			700	820	1180	1290	1460	1570	1680	1830	1970
		0.344	72.21	190.65			770	900	1300	1420	1610	1730	1860	2010	2170
		0.375	78.60	207.50	STD	20	840	980	1420	1550	1760	1890	2020	2190	2360
		0.406	84.96	224.30			910	1070	1530	1680	1900	2050	2190	2380	2560
		0.438	91.51	241.58			990	1150	1660	1810	2050	2210	2370	2560	2760
		0.469	97.83	258.27			1060	1230	1770	1940	2190	2360	2530	2740	2950
		0.500	104.13	274.90	XS	30	1120	1310	1890	2070	2340	2520	2700	2930	3000
		0.562	116.67	308.01			1260	1480	2120	2330	2630	2830	3000	3000	3000
		0.594	123.11	325.01		40	1340	1560	2250	2460	2930	3000	3000	3000	3000
		0.625	129.33	341.43			1410	1640	2360	2590	3000	3000	3000	3000	3000
		0.688	141.90	374.62			1550	1810	2600	2850	3000	3000	3000	3000	3000
		0.750	154.19	407.07			1690	1970	2840	3000	3000	3000	3000	3000	3000
		0.812	166.40	439.30		60	1830	2130	3000	3000	3000	3000	3000	3000	3000
		1.031	208.87	551.41		80	2320	2710	3000	3000	3000	3000	3000	3000	3000
		1.281	256.10	676.09		100	2800	3000	3000	3000	3000	3000	3000	3000	3000
		1.375	273.51	722.07			3000	3000	3000	3000	3000	3000	3000	3000	3000

USS Standard Pipe* and Line Pipe† Data (*Continued*)

Size		Nom. Wall Thick-ness, Inches	Weight Lb/Ft Plain End	Weight Tons/ Mile	Class	Sched. No.	Mill Hydrostatic Test Pressure‡ — psi								
NPS	OD Inches						Grade A	Grade B	Grade X-42	Grade X-46	Grade X-52	Grade X-56	Grade X-60	Grade X-65	Grade X-70
22	22.000 (Seamless)	0.281	65.18	172.08			570	670	970	1060	1200	1290	1380	1490	1610
		0.312	72.27	190.79			640	740	1070	1170	1330	1430	1530	1660	1790
		0.344	79.56	210.04			700	820	1180	1290	1460	1580	1690	1830	1970
		0.375	86.61	228.65	STD	20	770	890	1290	1410	1600	1720	1840	1990	2150
		0.406	93.63	247.19			830	970	1400	1530	1730	1860	1990	2160	2330
		0.438	100.86	266.28			900	1050	1510	1650	1860	2010	2150	2330	2510
		0.469	107.85	284.72			960	1120	1610	1770	2000	2150	2300	2490	2690
		0.500	114.81	303.10	XS	30	1020	1190	1720	1880	2130	2290	2450	2660	2860
		0.562	128.67	339.70			1150	1340	1930	2120	2390	2570	2760	2990	3000
		0.625	142.68	376.67			1280	1490	2150	2350	2660	2860	3000	3000	3000
		0.688	156.60	413.42			1410	1640	2360	2590	2930	3000	3000	3000	3000
		0.750	170.21	449.36			1530	1790	2580	2820	3000	3000	3000	3000	3000
		0.812	183.75	485.09			1660	1940	2790	3000	3000	3000	3000	3000	3000
		0.875	197.41	521.17		60	1790	2090	3000	3000	3000	3000	3000	3000	3000
		1.125	250.81	662.15		80	2300	2500	3000	3000	3000	3000	3000	3000	3000
		1.219	270.55	714.25			2490	2680	3000	3000	3000	3000	3000	3000	3000
		1.250	277.01	731.31			2560	2980	3000	3000	3000	3000	3000	3000	3000
24	24.000 (Seamless)	0.250	63.41	167.41			470	550	790	860	980	1050	1120	1220	1310
		0.281	71.18	187.92			530	610	890	970	1100	1180	1260	1370	1480
		0.312	78.93	208.38			580	680	980	1080	1220	1310	1400	1520	1640
		0.344	86.91	229.44			640	750	1080	1190	1340	1440	1550	1680	1810
		0.375	94.62	249.79	STD	20	700	820	1180	1290	1460	1580	1690	1830	1970
		0.406	102.31	270.09			760	890	1280	1400	1580	1710	1830	1980	2130
		0.438	110.22	290.98			820	960	1380	1510	1710	1840	1970	2140	2300
		0.469	117.86	311.16			880	1030	1480	1620	1830	1970	2110	2290	2460
		0.500	125.49	331.29	XS		940	1090	1580	1720	1950	2100	2250	2440	2630
		0.562	140.68	371.39		30	1050	1230	1770	1940	2190	2360	2530	2740	2950
		0.625	156.03	411.91			1170	1370	1970	2160	2440	2630	2810	3000	3000
		0.688	171.29	452.21		40	1290	1500	2170	2370	2680	2890	3000	3000	3000
		0.750	186.23	491.65			1410	1640	2360	2590	2930	3000	3000	3000	3000
		0.812	201.09	530.88			1520	1780	2560	2800	3000	3000	3000	3000	3000
		0.875	216.10	570.51			1640	1910	2760	3000	3000	3000	3000	3000	3000
		0.938	231.03	609.92			1760	2050	2950	3000	3000	3000	3000	3000	3000
		0.969	238.35	629.23		60	1820	2120	3000	3000	3000	3000	3000	3000	3000
		1.031	252.91	667.68			1930	2260	3000	3000	3000	3000	3000	3000	3000
		1.062	260.17	686.85			1990	2320	3000	3000	3000	3000	3000	3000	3000
		1.219	296.58	782.97		80	2290	2670	3000	3000	3000	3000	3000	3000	3000
		1.312	317.91	839.28			2460	2870	3000	3000	3000	3000	3000	3000	3000

USS Standard Pipe* and Line Pipe† Data (*Continued*)

Size		Nom. Wall Thick-ness, Inches	Weight Lb/Ft Plain End	Weight Tons/ Mile	Class	Sched. No.	Mill Hydrostatic Test Pressure‡ — psi								
NPS	OD Inches						Grade A	Grade B	Grade X-42	Grade X-46	Grade X-52	Grade X-56	Grade X-60	Grade X-65	Grade X-70
26	26.000 (Seamless)	0.250	68.75	181.51			430	500	730	800	900	970	1040	1120	1210
		0.281	77.18	203.77			490	570	820	890	1010	1090	1170	1260	1360
		0.312	85.60	225.97			540	630	910	990	1120	1210	1300	1400	1510
		0.344	94.26	248.84			600	690	1000	1100	1240	1330	1430	1550	1670
		0.375	102.63	270.94	STD		650	760	1090	1190	1350	1450	1560	1690	1820
		0.406	110.98	292.98			700	820	1180	1290	1460	1570	1690	1830	1970
		0.438	119.57	315.68			760	880	1270	1390	1580	1700	1820	1970	2120
		0.469	127.88	337.61			810	950	1360	1490	1690	1820	1950	2110	2270
		0.500	136.17	359.49	XS	20	870	1010	1450	1590	1800	1940	2080	2250	2420
		0.562	152.68	403.08			970	1130	1630	1790	2020	2180	2330	2530	2720
		0.625	169.38	447.16			1080	1260	1820	1990	2250	2420	2600	2810	3000
		0.688	185.99	491.01			1190	1390	2000	2190	2480	2670	2860	3000	3000
		0.750	202.25	533.95			1300	1510	2180	2390	2700	2910	3000	3000	3000
		0.812	218.43	576.66			1410	1640	2360	2590	2920	3000	3000	3000	3000
		0.938	251.07	662.82			1620	1890	2730	2990	3000	3000	3000	3000	3000
		1.062	282.85	746.72			1840	2140	3000	3000	3000	3000	3000	3000	3000
		1.188	314.81	831.10			2060	2400	3000	3000	3000	3000	3000	3000	3000

USS Standard Pipe and Line Pipe Comparative Specification Data

Specification	A53 NPS ⅛–26 STD, XS and XXS, ANSI Schedules 10 through 160
Scope	Covers Seamless and Welded, Black and hot-dipped galvanized nominal (average) wall pipe for coiling, bending, flanging and other special purposes and is suitable for welding. Purpose for which pipe is intended should be stated on order.
Kinds of Steel Permitted For Pipe Material	Open-hearth Basic-oxygen Electric-furnace
Hot-Dipped Galvanizing	Sets standards for coating of pipe with zinc inside and outside by the hot-dipped process. Weight of coating must not average less than 1.8 oz. per square foot and not less than 1.6 oz. per square foot.
Permissible Variations in Wall Thickness	The minimum wall thickness at any point shall not be more than 12.5% under the nominal wall thickness specified.
Chemical Requirements	C max % / Mn max % / P max % / S max % Seamless or ERW Grade A: 0.25 / 0.95 / 0.05 / 0.06 Grade B: 0.30 / 1.20 / 0.05 / 0.06
Tensile Requirements	Seamless and Electric-Resistance-Welded Grade A / Grade B Tensile Strength, min., psi............ 48,000 / 60,000 Yield Strength, min., psi............. 30,000 / 35,000
Hydrostatic Testing	Hydrostatic inspection test pressures for plain end and threaded and coupled pipe are specified. Hydrostatic pressure shall be maintained for not less than 5 seconds for all sizes of Seamless and Electric-Resistance-Weld Pipe.
Permissible Variations in Weights per Foot	Plus or Minus 10%
Permissible Variations in Outside Diameter	**Outside Diameter** at any point shall not vary from standard specified more than— **For NPS 1½ and Smaller Sizes**: 1/64" over, 1/32" under **For NPS 2 and Larger Sizes**: 1% over, 1% under
Mechanical Tests Specified	**Tensile Test**—Transverse required on ERW for NPS 8 and larger. **Bending Test (Cold)**—STD and XS-NPS 2 and under XXS-NPS 1¼ and under. Degree of Bend / Diameter of Mandrel For Normal A53 Uses: 90 / 12 x nom. dia. of pipe For Close Coiling: 180 / 8 x nom. dia. of pipe **Flattening Test**—NPS 2 and larger STD and XS. (Not required for XXS pipe).
Number of Tests Required	**Seamless and Electric-Resistance-Welded**—Bending, flattening, tensile on one length of pipe from each lot of 500 lengths or less of a size.
Lengths	**Standard Weight** **Single Random**—16'-22'. 5% may be jointers. If Plain Ends—5% may be 12'-16'. **Double Random**—Shortest Length 22', minimum average for order 35'. **Extra Strong & Double Extra Strong** **Single Random**—12'-22'. 5% may be 6'-12'. **Double Random (XS and lighter)**—Shortest Length 22', minimum average for order 35 . Lengths longer than single random with wall thicknesses heavier than XS subject to negotiation.
Required Markings on Each Length (On Tags attached to each Bundle in case of Bundled Pipe)	**Rolled, Stamped or Stenciled (Mfgrs. option)** Name or brand of manufacturer. Kind of pipe, that is, Electric-Resistance-Welded A, Electric-Resistance-Welded B, Seamless A; or Seamless B; XS for extra strong, XXS for double extra strong. ASTM A53. Length of pipe.
General Information	**Couplings**—Applied handling tight. Couplings, 2" and smaller straight tapped, other sizes taper tapped. **Thread Protection**—Applied to pipe 4" and larger. **End Finish** (unless otherwise specified)— STD or XS, or wall thicknesses less than 0.500 in. (excluding XXS): Plain end beveled. All XXS and wall thicknesses over 0.500 in.: Plain end square cut.

NOTE: This is summarized information from ASTM Standards and API Specification 5L. Please refer to the specific Standard or Specification for more details.

Specification	A106 NPS ⅛–48 ANSI Schedules to 160
Scope	Covers SEAMLESS carbon steel nominal wall pipe for high-temperature service, suitable for bending, flanging and similar forming operations. NPS 1½ and under may be either hot finished or cold drawn. NPS 2 and larger shall be hot finished unless otherwise specified.
Kinds of Steel Permitted For Pipe Material	Killed steel Open-hearth Electric-furnace Basic-oxygen
Hot-Dipped Galvanizing	Not covered in specification.
Permissible Variations in Wall Thickness	The minimum wall thickness at any point shall not be more than 12.5% under the nominal wall thickness specified.
Chemical Requirements	see table below
Tensile Requirements	see table below
Hydrostatic Testing	Inspection test pressures produce a stress in the pipe wall equal to 60% of specified minimum yield strength (SMYS) at room temperature. Maximum Pressures are not to exceed 2500 psi for NPS 3 and under, and 2800 psi for the larger sizes. Pressure is maintained for not less than 5 seconds.
Permissible Variations in Weights per Foot	Weight of any length shall not vary more than 10% over and 3.5% under that specified. NOTE — NPS 4 and smaller — weighed in lots. Larger sizes — by length.
Permissible Variations in Outside Diameter	Outside Diameter at any point shall not vary from standard specified more than— (see table below)
Mechanical Tests Specified	Tensile Test—NPS 8 and larger—either transverse or longitudinal acceptable. Smaller than NPS 8—longitudinal only. Flattening Test—NPS 2 and larger. Bending Test (Cold)—NPS 2 and under. (see table below)
Number of Tests Required	see table below
Lengths	Lengths required shall be specified on order. No "jointers" permitted unless otherwise specified. If no definite lengths required, following practice applies: Single Random—16′-22′. 5% may be 12′-16′. Double Random—Minimum length 22′, Minimum average 35′. 5% may be 16′-22′.
Required Markings On Each Length (On Tags attached to each Bundle in case of Bundled Pipe)	Rolled, Stamped or Stenciled (Mfgrs. option) Manufacturer's name or brand. A106 A, A106 B, or A106 C. Hydrostatic test pressure and/or NDE, or NH if neither is specified. Length of pipe. ANSI schedule number. Weight per foot (NPS 4 and larger). Additional "S" if tested to supplementary requirements.
General Information	• Unless otherwise specified, pipe furnished with plain ends. • Surface finish standards are outlined in specification. • Purchaser may specify NDE in lieu of hydrostatic test or neither.

Chemical Requirements

	Grade A	Grade B	Grade C
Carbon max. %	0.25	0.30	0.35
Manganese %	0.27 to 0.93	0.29 to 1.06	0.29 to 1.06
Phosphorus, max. %	0.035	0.035	0.035
Sulfur, max. %	0.035	0.035	0.035
Silicon, min. %	0.10	0.10	0.10

Tensile Requirements — Seamless

	Grade A	Grade B	Grade C
Tensile Strength, min., psi	48,000	60,000	70,000
Yield Strength, min., psi	30,000	35,000	40,000

Permissible Variations in Outside Diameter

NPS	Over	Under
1½ and smaller	1/64″	1/32″
2–4	1/32″	1/32″
5–8	1/16″	1/32″
10–18	3/32″	1/32″
20–26	⅛″	1/32″

Bending Test

	Degree of Bend	Diameter of Mandrel
For Normal A106 Uses	90	12 x nom. dia. of pipe
For Close Coiling	180	8 x nom. dia. of pipe

Number of Tests Required

	NPS	On One Length From Each Lot of
Tensile	5 and smaller	400 or less
	6 and larger	200 or less
Bending	2 and smaller	400 or less
Flattening	2 through 5	400 or less
	6 and over	200 or less

Appendix

B

Table of Boiler Tubes

Boiler tubes are most often installed into drums and headers by expanding through the use of tube-rolling tools. Watertubes are expanded and flared, while firetubes are expanded and beaded over. The tube holes must be larger than the OD in order to install the tubes. In boiler work, the clearance is typically about 1⁄32 inch on the diameter. After the tubes are installed, the rolling tool increases the tube diameter until the outside diameter of the tube contacts the walls of the hole in the tubesheet. Continued rolling from this point increases the inside diameter of the tube, and therefore reduces the thickness as well. (This is the reason for the tube-rolling allowance in PG-27.4, Note 4.) For carbon steel tubes, the optimum degree of rolling is given by Elliott as 7–8 percent thickness reduction. This is based on tubes having a before-rolling hardness in the range of Brinell 90 to 120. The chart below shows the amount of increase in the inside diameter of the tubes that corresponds to a 7½ percent thickness reduction of the tube wall. (This increase in ID is the increase after the tube has been expanded to contact the walls of the hole.) It can be seen that, in practice, measuring the ID changes on these boiler tubes would be impractical; the skill and experience of the boiler maker is the most important factor.

BWG no.	Thickness	I.D. increase for 7½% reduction in thickness
00	0.380	0.006
0	0.340	0.005
1	0.300	0.005
2	0.284	0.004
3	0.259	0.004
4	0.238	0.004
5	0.220	0.003
6	0.203	0.003
7	0.180	0.003
8	0.165	0.002
9	0.148	0.002
10	0.134	0.002
11	0.120	0.002
12	0.109	0.002
13	0.095	0.001
14	0.083	0.001
15	0.072	0.001
16	0.065	0.001
17	0.058	0.001
18	0.049	0.001

Appendix

C

Excerpt from Steam Tables

Saturation: Pressures—Table 2 from Keenan and Keyes

Abs. press. lb./sq. in.	Temp. Fahr.	Specific volume		Enthalpy			Entropy			Internal energy			Abs. press. lb./sq. in.
		Sat. liquid	Sat. vapor	Sat. liquid	Evap.	Sat. vapor	Sat. liquid	Evap.	Sat. vapor	Sat. liquid	Evap.	Sat. vapor	
p	t	v_f	v_g	h_f	h_{fg}	h_g	s_f	s_{fg}	s_g	u_f	u_{fg}	u_g	p
3.0	141.48	0.01630	118.71	109.37	1013.2	1122.6	0.2008	1.6855	1.8863	109.36	947.3	1056.7	3.0
3.5	147.57	0.01633	102.72	115.46	1009.6	1125.1	0.2109	1.6626	1.8735	115.45	943.1	1058.6	3.5
4.0	152.97	0.01636	90.63	120.86	1006.4	1127.3	0.2198	1.6427	1.8625	120.85	939.3	1060.2	4.0
4.5	157.83	0.01638	81.16	125.71	1003.6	1129.3	0.2276	1.6252	1.8528	125.70	936.0	1061.7	4.5
5.0	162.24	0.01640	73.52	130.13	1001.0	1131.1	0.2347	1.6094	1.8441	130.12	933.0	1063.1	5.0
5.5	166.30	0.01643	67.24	134.19	998.5	1132.7	0.2411	1.5951	1.8363	134.17	930.1	1064.3	5.5
6.0	170.06	0.01645	61.98	137.96	996.2	1134.2	0.2472	1.5820	1.8292	137.94	927.5	1065.4	6.0
6.5	173.56	0.01647	57.50	141.47	994.1	1135.6	0.2528	1.5699	1.8227	141.45	925.0	1066.4	6.5
7.0	176.85	0.01649	53.64	144.76	992.1	1136.9	0.2581	1.5586	1.8167	144.74	922.7	1067.4	7.0
7.5	179.94	0.01651	50.29	147.86	990.2	1138.1	0.2629	1.5481	1.8110	147.84	920.5	1068.3	7.5
8.0	182.86	0.01653	47.34	150.79	988.5	1139.3	0.2674	1.5383	1.8057	150.77	918.4	1069.2	8.0
8.5	185.64	0.01654	44.73	153.57	986.8	1140.4	0.2718	1.5290	1.8008	153.54	916.5	1070.0	8.5
9.0	188.28	0.01656	42.40	156.22	985.2	1141.4	0.2759	1.5203	1.7962	156.19	914.6	1070.8	9.0
9.5	190.80	0.01658	40.31	158.75	983.6	1142.3	0.2798	1.5120	1.7918	158.72	912.8	1071.5	9.5
10	193.21	0.01659	38.42	161.17	982.1	1143.3	0.2835	1.5041	1.7876	161.14	911.1	1072.2	10
11	197.75	0.01662	35.14	165.73	979.3	1145.0	0.2903	1.4897	1.7800	165.70	907.8	1073.5	11
12	201.96	0.01665	32.40	169.96	976.6	1146.6	0.2967	1.4763	1.7730	169.92	904.8	1074.7	12
13	205.88	0.01667	30.06	173.91	974.2	1148.1	0.3027	1.4638	1.7665	173.87	901.9	1075.8	13
14	209.56	0.01670	28.04	177.61	971.9	1149.5	0.3083	1.4522	1.7605	177.57	899.3	1076.9	14
14.696	212.00	0.01672	26.80	180.07	970.3	1150.4	0.3120	1.4446	1.7566	180.02	897.5	1077.5	14.696
15	213.03	0.01672	26.29	181.11	969.7	1150.8	0.3135	1.4415	1.7549	181.06	896.7	1077.8	15
16	216.32	0.01674	24.75	184.42	967.6	1152.0	0.3184	1.4313	1.7497	184.37	894.3	1078.7	16
17	219.44	0.01677	23.39	187.56	965.5	1153.1	0.3231	1.4218	1.7449	187.51	892.0	1079.5	17
18	222.41	0.01679	22.17	190.56	963.6	1154.2	0.3275	1.4128	1.7403	190.50	889.9	1080.4	18
19	225.24	0.01681	21.08	193.42	961.9	1155.3	0.3317	1.4043	1.7360	193.36	887.8	1081.2	19

20	227.96	0.01683	20.089	196.16	960.1	1156.3	0.3356	1.3962	1.7319	196.10	885.8	1081.9	20
21	230.57	0.01685	19.192	198.79	958.4	1157.2	0.3395	1.3885	1.7280	198.73	883.9	1082.6	21
22	233.07	0.01687	18.375	201.33	956.8	1158.1	0.3431	1.3811	1.7242	201.26	882.0	1083.3	22
23	235.49	0.01689	17.627	203.78	955.2	1159.0	0.3466	1.3740	1.7206	203.71	880.2	1083.9	23
24	237.82	0.01691	16.938	206.14	953.7	1159.8	0.3500	1.3672	1.7172	206.07	878.5	1084.6	24
25	240.07	0.01692	16.303	208.42	952.1	1160.6	0.3533	1.3606	1.7139	208.34	876.8	1085.1	25
26	242.25	0.01694	15.715	210.62	950.7	1161.3	0.3564	1.3544	1.7108	210.54	875.2	1085.7	26
27	244.36	0.01696	15.170	212.75	949.3	1162.0	0.3594	1.3484	1.7078	212.67	873.6	1086.3	27
28	246.41	0.01698	14.663	214.83	947.9	1162.7	0.3623	1.3425	1.7048	214.74	872.1	1086.8	28
29	248.40	0.01699	14.189	216.86	946.5	1163.4	0.3652	1.3368	1.7020	216.77	870.5	1087.3	29
30	250.33	0.01701	13.746	218.82	945.3	1164.1	0.3680	1.3313	1.6993	218.73	869.1	1087.8	30
31	252.22	0.01702	13.330	220.73	944.0	1164.7	0.3707	1.3260	1.6967	220.63	867.7	1088.3	31
32	254.05	0.01704	12.940	222.59	942.8	1165.4	0.3733	1.3209	1.6941	222.49	866.3	1088.7	32
33	255.84	0.01705	12.572	224.41	941.6	1166.0	0.3758	1.3159	1.6917	224.31	864.9	1089.2	33
34	257.58	0.01707	12.226	226.18	940.3	1166.5	0.3783	1.3110	1.6893	226.07	863.5	1089.6	34
35	259.28	0.01708	11.898	227.91	939.2	1167.1	0.3807	1.3063	1.6870	227.80	862.3	1090.1	35
36	260.95	0.01709	11.588	229.60	938.0	1167.6	0.3831	1.3017	1.6848	229.49	861.0	1090.5	36
37	262.57	0.01711	11.294	231.26	936.9	1168.2	0.3854	1.2972	1.6826	231.14	859.8	1090.9	37
38	264.16	0.01712	11.015	232.89	935.8	1168.7	0.3876	1.2929	1.6805	232.77	858.5	1091.3	38
39	265.72	0.01714	10.750	234.48	934.7	1169.2	0.3898	1.2886	1.6784	234.36	857.2	1091.6	39
40	267.25	0.01715	10.498	236.03	933.7	1169.7	0.3919	1.2844	1.6763	235.90	856.1	1092.0	40
41	268.74	0.01716	10.258	237.55	932.6	1170.2	0.3940	1.2803	1.6743	237.42	855.0	1092.4	41
42	270.21	0.01717	10.029	239.04	931.6	1170.7	0.3960	1.2764	1.6724	238.91	853.8	1092.7	42
43	271.64	0.01719	9.810	240.51	930.6	1171.1	0.3980	1.2726	1.6706	240.37	852.7	1093.1	43
44	273.05	0.01720	9.601	241.95	929.6	1171.6	0.4000	1.2687	1.6687	241.81	851.6	1093.4	44
45	274.44	0.01721	9.401	243.36	928.6	1172.0	0.4019	1.2650	1.6669	243.22	850.5	1093.7	45
46	275.80	0.01722	9.209	244.75	927.7	1172.4	0.4038	1.2613	1.6652	244.60	849.5	1094.1	46
47	277.13	0.01723	9.025	246.12	926.7	1172.9	0.4057	1.2577	1.6634	245.97	848.4	1094.4	47
48	278.45	0.01725	8.848	247.47	925.8	1173.3	0.4075	1.2542	1.6617	247.32	847.4	1094.7	48
49	279.74	0.01726	8.678	248.79	924.9	1173.7	0.4093	1.2508	1.6601	248.63	846.4	1095.0	49
50	281.01	0.01727	8.515	250.09	924.0	1174.1	0.4110	1.2474	1.6585	249.93	845.4	1095.3	50
51	282.26	0.01728	8.359	251.37	923.0	1174.4	0.4127	1.2442	1.6569	251.21	844.3	1095.5	51

Saturation: Pressures (*Continued*)

Abs. press. lb./sq. in.	Temp. Fahr.	Specific volume Sat. liquid	Specific volume Sat. vapor	Enthalpy Sat. liquid	Enthalpy Evap.	Enthalpy Sat. vapor	Entropy Sat. liquid	Entropy Evap.	Entropy Sat. vapor	Internal energy Sat. liquid	Internal energy Evap.	Internal energy Sat. vapor	Abs. press. lb./sq. in.
p	t	v_f	v_g	h_f	h_{fg}	h_g	s_f	s_{fg}	s_g	u_f	u_{fg}	u_g	p
52	283.49	0.01729	8.208	252.63	922.2	1174.8	0.4144	1.2409	1.6553	252.46	843.3	1095.8	52
53	284.70	0.01730	8.062	253.87	921.3	1175.2	0.4161	1.2377	1.6538	253.70	842.4	1096.1	53
54	285.90	0.01731	7.922	255.09	920.5	1175.6	0.4177	1.2346	1.6523	254.92	841.5	1096.4	54
55	287.07	0.01732	7.787	256.30	919.6	1175.9	0.4193	1.2316	1.6509	256.12	840.6	1096.7	55
56	288.23	0.01733	7.656	257.50	918.8	1176.3	0.4209	1.2285	1.6494	257.32	839.7	1097.0	56
57	289.37	0.01734	7.529	258.67	917.9	1176.6	0.4225	1.2255	1.6480	258.49	838.7	1097.2	57
58	290.50	0.01736	7.407	259.82	917.1	1176.9	0.4240	1.2226	1.6466	259.63	837.8	1097.4	58
59	291.61	0.01737	7.289	260.96	916.3	1177.3	0.4255	1.2197	1.6452	260.77	836.9	1097.7	59
60	292.71	0.01738	7.175	262.09	915.5	1177.6	0.4270	1.2168	1.6438	261.90	836.0	1097.9	60
61	293.79	0.01739	7.064	263.20	914.7	1177.9	0.4285	1.2140	1.6425	263.00	835.2	1098.2	61
62	294.85	0.01740	6.957	264.30	913.9	1178.2	0.4300	1.2112	1.6412	264.10	834.3	1098.4	62
63	295.90	0.01741	6.853	265.38	913.1	1178.5	0.4314	1.2085	1.6399	265.18	833.4	1098.6	63
64	296.94	0.01742	6.752	266.45	912.3	1178.8	0.4328	1.2059	1.6387	266.24	832.6	1098.8	64
65	297.97	0.01743	6.655	267.50	911.6	1179.1	0.4342	1.2032	1.6374	267.29	831.8	1099.1	65
66	298.99	0.01744	6.560	268.55	910.8	1179.4	0.4356	1.2006	1.6362	268.34	831.0	1099.3	66
67	299.99	0.01745	6.468	269.58	910.1	1179.7	0.4369	1.1981	1.6350	269.36	830.2	1099.5	67
68	300.98	0.01746	6.378	270.60	909.4	1180.0	0.4383	1.1955	1.6338	270.38	829.4	1099.8	68
69	301.96	0.01747	6.291	271.61	908.7	1180.3	0.4396	1.1930	1.6326	271.39	828.6	1100.0	69
70	302.92	0.01748	6.206	272.61	907.9	1180.6	0.4409	1.1906	1.6315	272.38	827.8	1100.2	70
71	303.88	0.01749	6.124	273.60	907.2	1180.8	0.4422	1.1881	1.6303	273.37	827.0	1100.4	71
72	304.83	0.01750	6.044	274.57	906.5	1181.1	0.4435	1.1857	1.6292	274.34	826.3	1100.6	72
73	305.76	0.01751	5.966	275.54	905.8	1181.3	0.4447	1.1834	1.6281	275.30	825.5	1100.8	73
74	306.68	0.01752	5.890	276.49	905.1	1181.6	0.4460	1.1810	1.6270	276.25	824.7	1101.0	74
75	307.60	0.01753	5.816	277.43	904.5	1181.9	0.4472	1.1787	1.6259	277.19	824.0	1101.2	75
76	308.50	0.01754	5.743	278.37	903.7	1182.1	0.4484	1.1764	1.6248	278.12	823.3	1101.4	76

77	309.40	0.01754	5.673	279.30	903.1	1182.4	0.4496	1.1742	1.6238	279.05	822.5	1101.6	77
78	310.29	0.01755	5.604	280.21	902.4	1182.6	0.4508	1.1720	1.6228	279.96	821.7	1101.7	78
79	311.16	0.01756	5.537	281.12	901.7	1182.8	0.4520	1.1698	1.6217	280.86	821.0	1101.9	79
80	312.03	0.01757	5.472	282.02	901.1	1183.1	0.4531	1.1676	1.6207	281.76	820.3	1102.1	80
81	312.89	0.01758	5.408	282.91	900.4	1183.3	0.4543	1.1654	1.6197	282.65	819.6	1102.2	81
82	313.74	0.01759	5.346	283.79	899.7	1183.5	0.4554	1.1633	1.6187	283.52	818.9	1102.4	82
83	314.59	0.01760	5.285	284.66	899.1	1183.8	0.4565	1.1612	1.6177	284.39	818.2	1102.6	83
84	315.42	0.01761	5.226	285.53	898.5	1184.0	0.4576	1.1592	1.6168	285.26	817.5	1102.8	84
85	316.25	0.01761	5.168	286.39	897.8	1184.2	0.4587	1.1571	1.6158	286.11	816.8	1102.9	85
86	317.07	0.01762	5.111	287.24	897.2	1184.4	0.4598	1.1551	1.6149	286.96	816.1	1103.1	86
87	317.88	0.01763	5.055	288.08	896.5	1184.6	0.4609	1.1530	1.6139	287.80	815.4	1103.2	87
88	318.68	0.01764	5.001	288.91	895.9	1184.8	0.4620	1.1510	1.6130	288.63	814.8	1103.4	88
89	319.48	0.01765	4.948	289.74	895.3	1185.1	0.4630	1.1491	1.6121	289.45	814.1	1103.6	89
90	320.27	0.01766	4.896	290.56	894.7	1185.3	0.4641	1.1471	1.6112	290.27	813.4	1103.7	90
91	321.06	0.01767	4.845	291.38	894.1	1185.5	0.4651	1.1452	1.6103	291.08	812.8	1103.9	91
92	321.83	0.01768	4.796	292.18	893.5	1185.7	0.4661	1.1433	1.6094	291.88	812.2	1104.1	92
93	322.60	0.01768	4.747	292.98	892.9	1185.9	0.4672	1.1413	1.6085	292.68	811.5	1104.2	93
94	323.36	0.01769	4.699	293.78	892.3	1186.1	0.4682	1.1394	1.6076	293.47	810.9	1104.4	94
95	324.12	0.01770	4.652	294.56	891.7	1186.2	0.4692	1.1376	1.6068	294.25	810.2	1104.5	95
96	324.87	0.01771	4.606	295.34	891.1	1186.4	0.4702	1.1358	1.6060	295.03	809.6	1104.6	96
97	325.61	0.01772	4.561	296.12	890.5	1186.6	0.4711	1.1340	1.6051	295.80	808.9	1104.7	97
98	326.35	0.01772	4.517	296.89	889.9	1186.8	0.4721	1.1322	1.6043	296.57	808.3	1104.9	98
99	327.08	0.01773	4.474	297.65	889.4	1187.0	0.4731	1.1304	1.6035	297.33	807.7	1105.0	99
100	327.81	0.01774	4.432	298.40	888.8	1187.2	0.4740	1.1286	1.6026	298.08	807.1	1105.2	100
101	328.53	0.01775	4.391	299.15	888.2	1187.4	0.4750	1.1268	1.6018	298.82	806.5	1105.3	101
102	329.25	0.01775	4.350	299.90	887.6	1187.5	0.4759	1.1251	1.6010	299.57	805.9	1105.4	102
103	329.96	0.01776	4.310	300.64	887.1	1187.7	0.4768	1.1234	1.6002	300.30	805.3	1105.6	103
104	330.66	0.01777	4.271	301.37	886.5	1187.9	0.4778	1.1216	1.5994	301.03	804.7	1105.7	104
105	331.36	0.01778	4.232	302.10	886.0	1188.1	0.4787	1.1199	1.5986	301.75	804.1	1105.9	105
106	332.05	0.01778	4.194	302.82	885.4	1188.2	0.4796	1.1182	1.5978	302.47	803.5	1106.0	106
107	332.74	0.01779	4.157	303.54	884.9	1188.4	0.4805	1.1166	1.5971	303.19	802.9	1106.1	107
108	333.42	0.01780	4.120	304.26	884.3	1188.6	0.4814	1.1149	1.5963	303.90	802.4	1106.3	108
109	334.10	0.01781	4.084	304.97	883.7	1188.7	0.4823	1.1133	1.5956	304.61	801.8	1106.4	109

Saturation: Pressures (*Continued*)

		Specific volume		Enthalpy			Entropy			Internal energy			
Abs. press. lb./sq. in.	Temp. Fahr.	Sat. liquid	Sat. vapor	Sat. liquid	Evap.	Sat. vapor	Sat. liquid	Evap.	Sat. vapor	Sat. liquid	Evap.	Sat. vapor	Abs. press. lb./sq. in.
p	t	v_f	v_g	h_f	h_{fg}	h_g	s_f	s_{fg}	s_g	u_f	u_{fg}	u_g	p
110	334.77	0.01782	4.049	305.66	883.2	1188.9	0.4832	1.1117	1.5948	305.30	801.2	1106.5	110
111	335.44	0.01782	4.015	306.37	882.6	1189.0	0.4840	1.1101	1.5941	306.00	800.6	1106.6	111
112	336.11	0.01783	3.981	307.06	882.1	1189.2	0.4849	1.1085	1.5934	306.69	800.0	1106.7	112
113	336.77	0.01784	3.947	307.75	881.6	1189.4	0.4858	1.1069	1.5927	307.38	799.4	1106.8	113
114	337.42	0.01784	3.914	308.43	881.1	1189.5	0.4866	1.1053	1.5919	308.05	798.9	1106.9	114
115	338.07	0.01785	3.882	309.11	880.6	1189.7	0.4875	1.1037	1.5912	308.73	798.4	1107.1	115
116	338.72	0.01786	3.850	309.79	880.0	1189.8	0.4883	1.1022	1.5905	309.41	797.8	1107.2	116
117	339.36	0.01787	3.819	310.46	879.5	1190.0	0.4891	1.1007	1.5898	310.07	797.2	1107.3	117
118	339.99	0.01787	3.788	311.12	879.0	1190.1	0.4900	1.0992	1.5891	310.73	796.7	1107.4	118
119	340.62	0.01788	3.758	311.78	878.4	1190.2	0.4908	1.0977	1.5885	311.39	796.1	1107.5	119
120	341.25	0.01789	3.728	312.44	877.9	1190.4	0.4916	1.0962	1.5878	312.05	795.6	1107.6	120
121	341.88	0.01790	3.699	313.10	877.4	1190.5	0.4924	1.0947	1.5871	312.70	795.0	1107.7	121
122	342.50	0.01791	3.670	313.75	876.9	1190.7	0.4932	1.0933	1.5865	313.35	794.5	1107.8	122
123	343.11	0.01791	3.642	314.40	876.4	1190.8	0.4940	1.0918	1.5858	313.99	793.9	1107.9	123
124	343.72	0.01792	3.614	315.04	875.9	1190.9	0.4948	1.0903	1.5851	314.63	793.4	1108.0	124
125	344.33	0.01792	3.587	315.68	875.4	1191.1	0.4956	1.0888	1.5844	315.26	792.8	1108.1	125
126	344.94	0.01793	3.560	316.31	874.9	1191.2	0.4964	1.0874	1.5838	315.89	792.3	1108.2	126
127	345.54	0.01794	3.533	316.94	874.4	1191.3	0.4972	1.0859	1.5831	316.52	791.8	1108.3	127
128	346.13	0.01794	3.507	317.57	873.9	1191.5	0.4980	1.0845	1.5825	317.15	791.3	1108.4	128
129	346.73	0.01795	3.481	318.19	873.4	1191.6	0.4987	1.0832	1.5819	317.77	790.7	1108.5	129
130	347.32	0.01796	3.455	318.81	872.9	1191.7	0.4995	1.0817	1.5812	318.38	790.2	1108.6	130
131	347.90	0.01797	3.430	319.43	872.5	1191.9	0.5002	1.0804	1.5806	318.99	789.7	1108.7	131
132	348.48	0.01797	3.405	320.04	872.0	1192.0	0.5010	1.0790	1.5800	319.60	789.2	1108.8	132
133	349.06	0.01798	3.381	320.65	871.5	1192.1	0.5018	1.0776	1.5793	320.21	788.7	1108.9	133
134	349.64	0.01799	3.357	321.25	871.0	1192.2	0.5025	1.0762	1.5787	320.80	788.2	1109.0	134

135	350.21	0.01800	3.333	321.85	870.6	1192.4	0.5032	1.0749	1.5781	321.40	787.7	1109.1	135
136	350.78	0.01800	3.310	322.45	870.1	1192.5	0.5040	1.0735	1.5775	322.00	787.2	1109.2	136
137	351.35	0.01801	3.287	323.05	869.6	1192.6	0.5047	1.0722	1.5769	322.59	786.7	1109.3	137
138	351.91	0.01801	3.264	323.64	869.1	1192.7	0.5054	1.0709	1.5763	323.18	786.2	1109.4	138
139	352.47	0.01802	3.242	324.23	868.7	1192.9	0.5061	1.0696	1.5757	323.77	785.7	1109.5	139
140	353.02	0.01802	3.220	324.82	868.2	1193.0	0.5069	1.0682	1.5751	324.35	785.2	1109.6	140
141	353.57	0.01803	3.198	325.40	867.7	1193.1	0.5076	1.0669	1.5745	324.93	784.8	1109.7	141
142	354.12	0.01804	3.177	325.98	867.2	1193.2	0.5083	1.0657	1.5740	325.51	784.3	1109.8	142
143	354.67	0.01804	3.155	326.56	866.7	1193.3	0.5090	1.0644	1.5734	326.08	783.8	1109.8	143
144	355.21	0.01805	3.134	327.13	866.3	1193.4	0.5097	1.0631	1.5728	326.65	783.3	1109.9	144
145	355.76	0.01806	3.114	327.70	865.8	1193.5	0.5104	1.0618	1.5722	327.22	782.8	1110.0	145
146	356.29	0.01806	3.094	328.27	865.3	1193.6	0.5111	1.0605	1.5716	327.78	782.3	1110.1	146
147	356.83	0.01807	3.074	328.83	864.9	1193.8	0.5118	1.0592	1.5710	328.34	781.9	1110.2	147
148	357.36	0.01808	3.054	329.39	864.5	1193.9	0.5124	1.0580	1.5705	328.90	781.4	1110.3	148
149	357.89	0.01808	3.034	329.95	864.0	1194.0	0.5131	1.0568	1.5699	329.45	780.9	1110.4	149
150	358.42	0.01809	3.015	330.51	863.6	1194.1	0.5138	1.0556	1.5694	330.01	780.5	1110.5	150
152	359.46	0.01810	2.977	331.61	862.7	1194.3	0.5151	1.0532	1.5683	331.10	779.5	1110.6	152
154	360.49	0.01812	2.940	332.70	861.8	1194.5	0.5165	1.0507	1.5672	332.18	778.5	1110.7	154
156	361.52	0.01813	2.904	333.79	860.9	1194.7	0.5178	1.0483	1.5661	333.26	777.6	1110.9	156
158	362.53	0.01814	2.869	334.86	860.0	1194.9	0.5191	1.0459	1.5650	334.23	776.8	1111.0	158
160	363.53	0.01815	2.834	335.93	859.2	1195.1	0.5204	1.0436	1.5640	335.39	775.8	1111.2	160
162	364.53	0.01817	2.801	336.98	858.3	1195.3	0.5216	1.0414	1.5630	336.44	775.0	1111.4	162
164	365.51	0.01818	2.768	338.02	857.5	1195.5	0.5229	1.0391	1.5620	337.47	774.1	1111.5	164
166	366.48	0.01819	2.736	339.05	856.6	1195.7	0.5241	1.0369	1.5610	338.49	773.2	1111.7	166
168	367.45	0.01820	2.705	340.07	855.7	1195.8	0.5254	1.0346	1.5600	339.51	772.3	1111.8	168
170	368.41	0.01822	2.675	341.09	854.9	1196.0	0.5266	1.0324	1.5590	340.52	771.4	1111.9	170
172	369.35	0.01823	2.645	342.10	854.1	1196.2	0.5278	1.0302	1.5580	341.52	770.5	1112.0	172
174	370.29	0.01824	2.616	343.10	853.3	1196.4	0.5290	1.0280	1.5570	342.51	769.7	1112.2	174
176	371.22	0.01825	2.587	344.09	852.4	1196.5	0.5302	1.0259	1.5561	343.50	768.8	1112.3	176
178	372.14	0.01826	2.559	345.06	851.6	1196.7	0.5313	1.0238	1.5551	344.46	767.9	1112.4	178
180	373.06	0.01827	2.532	346.03	850.8	1196.9	0.5325	1.0217	1.5542	345.42	767.1	1112.5	180
182	373.96	0.01829	2.505	347.00	850.0	1197.0	0.5336	1.0196	1.5532	346.38	766.2	1112.6	182

Saturation: Pressures (*Continued*)

		Specific volume		Enthalpy			Entropy			Internal energy			
Abs. press. lb./sq. in.	Temp. Fahr.	Sat. liquid	Sat. vapor	Sat. liquid	Evap.	Sat. vapor	Sat. liquid	Evap.	Sat. vapor	Sat. liquid	Evap.	Sat. vapor	Abs. press. lb./sq. in.
p	t	v_f	v_g	h_f	h_{fg}	h_g	s_f	s_{fg}	s_g	u_f	u_{fg}	u_g	p
184	374.86	0.01830	2.479	347.96	849.2	1197.2	0.5348	1.0175	1.5523	347.34	765.4	1112.8	184
186	375.75	0.01831	2.454	348.92	848.4	1197.3	0.5359	1.0155	1.5514	348.29	764.6	1112.9	186
188	376.64	0.01832	2.429	349.86	847.6	1197.5	0.5370	1.0136	1.5506	349.22	763.8	1113.0	188
190	377.51	0.01833	2.404	350.79	846.8	1197.6	0.5381	1.0116	1.5497	350.15	763.0	1113.1	190
192	378.38	0.01834	2.380	351.72	846.1	1197.8	0.5392	1.0096	1.5488	351.07	762.1	1113.2	192
194	379.24	0.01835	2.356	352.64	845.3	1197.9	0.5403	1.0076	1.5479	351.98	761.3	1113.3	194
196	380.10	0.01836	2.333	353.55	844.5	1198.1	0.5414	1.0056	1.5470	352.89	760.6	1113.5	196
198	380.95	0.01838	2.310	354.46	843.7	1198.2	0.5425	1.0037	1.5462	353.79	759.8	1113.6	198
200	381.79	0.01839	2.288	355.36	843.0	1198.4	0.5435	1.0018	1.5453	354.68	759.0	1113.7	200
205	383.86	0.01842	2.234	357.58	841.1	1198.7	0.5461	0.9971	1.5432	356.88	757.1	1114.0	205
210	385.90	0.01844	2.183	359.77	839.2	1199.0	0.5487	0.9925	1.5412	359.05	755.2	1114.2	210
215	387.89	0.01847	2.134	361.91	837.4	1199.3	0.5512	0.9880	1.5392	361.18	753.2	1114.4	215
220	389.86	0.01850	2.087	364.02	835.6	1199.6	0.5537	0.9835	1.5372	363.27	751.3	1114.6	220
225	391.79	0.01852	2.0422	366.09	833.8	1199.9	0.5561	0.9792	1.5353	365.32	749.5	1114.8	225
230	393.68	0.01854	1.9992	368.13	832.0	1200.1	0.5585	0.9750	1.5334	367.34	747.7	1115.0	230
235	395.54	0.01857	1.9579	370.14	830.3	1200.4	0.5608	0.9708	1.5316	369.33	745.9	1115.3	235
240	397.37	0.01860	1.9183	372.12	828.5	1200.6	0.5631	0.9667	1.5298	371.29	744.1	1115.4	240
245	399.18	0.01863	1.8803	374.08	826.8	1200.9	0.5653	0.9627	1.5280	373.23	742.4	1115.6	245
250	400.95	0.01865	1.8438	376.00	825.1	1201.1	0.5675	0.9588	1.5263	375.14	740.7	1115.8	250
255	402.70	0.01868	1.8086	377.89	823.4	1201.3	0.5697	0.9549	1.5246	377.01	739.0	1116.0	255
260	404.42	0.01870	1.7748	379.76	821.8	1201.5	0.5719	0.9510	1.5229	378.86	737.3	1116.1	260
265	406.11	0.01873	1.7422	381.60	820.1	1201.7	0.5740	0.9472	1.5212	380.68	735.6	1116.3	265
270	407.78	0.01875	1.7107	383.42	818.5	1201.9	0.5760	0.9436	1.5196	382.48	733.9	1116.4	270

275	409.43	0.01878	1.6804	385.21	816.9	1202.1	0.5781	0.9399	1.5180	384.26	732.3	1116.6	275
280	411.05	0.01880	1.6511	386.98	815.3	1202.3	0.5801	0.9363	1.5164	386.01	730.7	1116.7	280
285	412.65	0.01883	1.6228	388.73	813.7	1202.4	0.5821	0.9327	1.5149	387.74	729.1	1116.8	285
290	414.23	0.01885	1.5954	390.46	812.1	1202.6	0.5841	0.9292	1.5133	389.45	727.5	1116.9	290
295	415.79	0.01887	1.5689	392.16	810.5	1202.7	0.5860	0.9258	1.5118	391.13	725.9	1117.0	295
300	417.33	0.01890	1.5433	393.84	809.0	1202.8	0.5879	0.9225	1.5104	392.79	724.3	1117.1	300
310	420.35	0.01894	1.4944	397.15	806.0	1203.1	0.5916	0.9159	1.5075	396.06	721.3	1117.4	310
320	423.29	0.01899	1.4485	400.39	803.0	1203.4	0.5952	0.9094	1.5046	399.26	718.3	1117.6	320
330	426.16	0.01904	1.4053	403.56	800.0	1203.6	0.5988	0.9031	1.5019	402.40	715.4	1117.8	330
340	428.97	0.01908	1.3645	406.66	797.1	1203.7	0.6022	0.8970	1.4992	405.46	712.4	1117.9	340
350	431.72	0.01913	1.3260	409.69	794.2	1203.9	0.6056	0.8910	1.4966	408.45	709.6	1118.0	350
360	434.40	0.01917	1.2895	412.67	791.4	1204.1	0.6090	0.8851	1.4941	411.39	706.8	1118.2	360
370	437.03	0.01921	1.2550	415.59	788.6	1204.2	0.6122	0.8794	1.4916	414.27	704.0	1118.3	370
380	439.60	0.01925	1.2222	418.45	785.8	1204.3	0.6153	0.8738	1.4891	417.10	701.3	1118.4	380
390	442.12	0.01930	1.1910	421.27	783.1	1204.4	0.6184	0.8683	1.4867	419.88	698.6	1118.5	390
400	444.59	0.0193	1.1613	424.0	780.5	1204.5	0.6214	0.8630	1.4844	422.6	695.9	1118.5	400
410	447.01	0.0194	1.1330	426.8	777.7	1204.5	0.6243	0.8578	1.4821	425.3	693.3	1118.6	410
420	449.39	0.0194	1.1061	429.4	775.2	1204.6	0.6272	0.8527	1.4799	427.9	690.8	1118.7	420
430	451.73	0.0194	1.0803	432.1	772.5	1204.6	0.6301	0.8476	1.4777	430.5	688.2	1118.7	430
440	454.02	0.0195	1.0556	434.6	770.0	1204.6	0.6329	0.8426	1.4755	433.0	685.7	1118.7	440
450	456.28	0.0195	1.0320	437.2	767.4	1204.6	0.6356	0.8378	1.4734	435.5	683.2	1118.7	450
460	458.50	0.0196	1.0094	439.7	764.9	1204.6	0.6383	0.8330	1.4713	438.0	680.7	1118.7	460
470	460.68	0.0196	0.9878	442.2	762.4	1204.6	0.6410	0.8283	1.4693	440.5	678.2	1118.7	470
480	462.82	0.0197	0.9670	444.6	759.9	1204.5	0.6436	0.8237	1.4673	442.9	675.7	1118.6	480
490	464.93	0.0197	0.9470	447.0	757.5	1204.5	0.6462	0.8191	1.4653	445.2	673.4	1118.6	490

Saturation: Pressures (*Continued*)

		Specific volume			Enthalpy			Entropy			Internal energy		
Abs. press. lb./sq. in.	Temp. Fahr.	Sat. liquid	Evap.	Sat. vapor	Sat. liquid	Evap.	Sat. vapor	Sat. liquid	Evap.	Sat. vapor	Sat. liquid	Sat. vapor	Abs. press. lb./sq. in.
p	t	v_f	v_{fg}	v_g	h_f	h_{fg}	h_g	s_f	s_{fg}	s_g	u_f	u_g	p
500	467.01	0.0197	0.9081	0.9278	449.4	755.0	1204.4	0.6487	0.8147	1.4634	447.6	1118.6	500
520	471.07	0.0198	0.8717	0.8915	454.1	750.1	1204.2	0.6536	0.8060	1.4596	452.2	1118.4	520
540	475.01	0.0199	0.8379	0.8578	458.6	745.4	1204.0	0.6584	0.7976	1.4560	456.6	1118.3	540
560	478.85	0.0200	0.8065	0.8265	463.0	740.8	1203.8	0.6631	0.7893	1.4524	460.9	1118.2	560
580	482.58	0.0201	0.7772	0.7973	467.4	736.1	1203.5	0.6676	0.7813	1.4489	465.2	1118.0	580
600	486.21	0.0201	0.7497	0.7698	471.6	731.6	1203.2	0.6720	0.7734	1.4454	469.4	1117.7	600
620	489.75	0.0202	0.7238	0.7440	475.7	727.2	1202.9	0.6763	0.7658	1.4421	473.4	1117.5	620
640	493.21	0.0203	0.6995	0.7198	479.8	722.7	1202.5	0.6805	0.7584	1.4389	477.4	1117.3	640
660	496.58	0.0204	0.6767	0.6971	483.8	718.3	1202.1	0.6846	0.7512	1.4358	481.3	1117.0	660
680	499.88	0.0204	0.6553	0.6757	487.7	714.0	1201.7	0.6886	0.7441	1.4327	485.1	1116.7	680
700	503.10	0.0205	0.6349	0.6554	491.5	709.7	1201.2	0.6925	0.7371	1.4296	488.8	1116.3	700
720	506.25	0.0206	0.6156	0.6362	495.3	705.4	1200.7	0.6963	0.7303	1.4266	492.5	1116.0	720
740	509.34	0.0207	0.5973	0.6180	499.0	701.2	1200.2	0.7001	0.7237	1.4237	496.2	1115.6	740
760	512.36	0.0207	0.5800	0.6007	502.6	697.1	1199.7	0.7037	0.7172	1.4209	499.7	1115.2	760
780	515.33	0.0208	0.5635	0.5843	506.2	692.9	1199.1	0.7073	0.7108	1.4181	503.2	1114.8	780
800	518.23	0.0209	0.5478	0.5687	509.7	688.9	1198.6	0.7108	0.7045	1.4153	506.6	1114.4	800
820	521.08	0.0209	0.5329	0.5538	513.2	684.8	1198.0	0.7143	0.6983	1.4126	510.0	1114.0	820
840	523.88	0.0210	0.5186	0.5396	516.6	680.8	1197.4	0.7177	0.6922	1.4099	513.3	1113.6	840
860	526.63	0.0211	0.5049	0.5260	520.0	676.8	1196.8	0.7210	0.6862	1.4072	516.6	1113.1	860
880	529.33	0.0212	0.4918	0.5130	523.3	672.8	1196.1	0.7243	0.6803	1.4046	519.9	1112.6	880
900	531.98	0.0212	0.4794	0.5006	526.6	668.8	1195.4	0.7275	0.6744	1.4020	523.1	1112.1	900
920	534.59	0.0213	0.4673	0.4886	529.8	664.9	1194.7	0.7307	0.6687	1.3995	526.2	1111.5	920
940	537.16	0.0214	0.4558	0.4772	533.0	661.0	1194.0	0.7339	0.6631	1.3970	529.3	1111.0	940
960	539.68	0.0214	0.4449	0.4663	536.2	657.1	1193.3	0.7370	0.6576	1.3945	532.4	1110.5	960
980	542.17	0.0215	0.4342	0.4557	539.3	653.3	1192.6	0.7400	0.6521	1.3921	535.4	1110.0	980

1000	544.61	0.0216	0.4240	0.4456	542.4	649.4	1191.8	0.7430	0.6467	1.3897	538.4	1109.4	1000
1050	550.57	0.0218	0.4000	0.4218	550.0	639.9	1189.9	0.7504	0.6334	1.3838	545.8	1108.0	1050
1100	556.31	0.0220	0.3781	0.4001	557.4	630.4	1187.8	0.7575	0.6205	1.3780	552.9	1106.4	1100
1150	561.86	0.0221	0.3581	0.3802	564.6	621.0	1185.6	0.7644	0.6079	1.3723	559.9	1104.7	1150
1200	567.22	0.0223	0.3396	0.3619	571.7	611.7	1183.4	0.7711	0.5956	1.3667	566.7	1103.0	1200
1250	572.42	0.0225	0.3225	0.3450	578.6	602.4	1181.0	0.7776	0.5836	1.3612	573.4	1101.2	1250
1300	577.46	0.0227	0.3066	0.3293	585.4	593.2	1178.6	0.7840	0.5719	1.3559	580.0	1099.4	1300
1350	582.35	0.0229	0.2919	0.3148	592.1	584.0	1176.1	0.7902	0.5604	1.3506	586.4	1097.5	1350
1400	587.10	0.0231	0.2781	0.3012	598.7	574.7	1173.4	0.7963	0.5491	1.3454	592.7	1095.4	1400
1450	591.73	0.0233	0.2651	0.2884	605.2	565.5	1170.7	0.8023	0.5379	1.3402	599.0	1093.3	1450
1500	596.23	0.0235	0.2530	0.2765	611.6	556.3	1167.9	0.8082	0.5269	1.3351	605.1	1091.2	1500
1600	604.90	0.0239	0.2309	0.2548	624.1	538.0	1162.1	0.8196	0.5053	1.3249	617.0	1086.7	1600
1700	613.15	0.0243	0.2111	0.2354	636.3	519.6	1155.9	0.8306	0.4843	1.3149	628.7	1081.8	1700
1800	621.03	0.0247	0.1932	0.2179	648.3	501.1	1149.4	0.8412	0.4637	1.3049	640.1	1076.8	1800
1900	628.58	0.0252	0.1769	0.2021	660.1	482.4	1142.4	0.8516	0.4433	1.2949	651.2	1071.4	1900
2000	635.82	0.0257	0.1621	0.1878	671.7	463.4	1135.1	0.8619	0.4230	1.2849	662.2	1065.6	2000
2100	642.77	0.0262	0.1484	0.1746	683.3	444.1	1127.4	0.8721	0.4027	1.2748	673.1	1059.6	2100
2200	649.46	0.0268	0.1358	0.1625	694.8	424.4	1119.2	0.8820	0.3826	1.2646	683.9	1053.1	2200
2300	655.91	0.0274	0.1239	0.1513	706.5	403.9	1110.4	0.8921	0.3621	1.2541	694.8	1046.0	2300
2400	662.12	0.0280	0.1128	0.1407	718.4	382.7	1101.1	0.9023	0.3411	1.2434	706.0	1038.6	2400
2500	668.13	0.0287	0.1021	0.1307	730.6	360.5	1091.1	0.9126	0.3197	1.2322	717.3	1030.6	2500
2600	673.94	0.0295	0.0918	0.1213	743.0	337.2	1080.2	0.9232	0.2973	1.2205	728.8	1021.9	2600
2700	679.55	0.0305	0.0818	0.1123	756.2	312.1	1068.3	0.9342	0.2740	1.2082	741.0	1012.3	2700
2800	684.99	0.0315	0.0719	0.1035	770.1	284.7	1054.8	0.9459	0.2487	1.1946	753.8	1001.2	2800
2900	690.26	0.0329	0.0618	0.0947	785.4	253.6	1039.0	0.9587	0.2205	1.1792	767.7	988.2	2900
3000	695.36	0.0346	0.0512	0.0858	802.5	217.8	1020.3	0.9731	0.1885	1.1615	783.4	972.7	3000
3100	700.31	0.0371	0.0382	0.0753	825.0	168.1	993.1	0.9919	0.1449	1.1368	803.7	949.9	3100
3200	705.11	0.0444	0.0136	0.0580	872.4	62.0	934.4	1.0320	0.0532	1.0852	846.0	898.4	3200
3206.2	705.40	0.0503	0	0.0503	902.7	0	902.7	1.0580	0	1.0580	872.9	872.9	3206.2

Reprinted by permission of Keenan and Keyes.

Appendix

D

Table of Units and Conversions to SI Units

The 1995 edition of the Boiler and Pressure Vessel Code is based on U.S. Customary (ft-lb) units of measurement, which are to be regarded as the standard. This supplement is provided as a convenience to the Code user and contains SI conversion factors for units contained in the Code.

List of SI Units for Use with ASME Boiler and Pressure Vessel Code[1]

Quantity	Unit	Symbol	Other Units or Limitations
Space and Time			
plane angle	radian	rad	degree (decimalized)
length	meter	m	
area	square meter	m^2	
volume	cubic meter	m^3	liter (L) for liquid only (use without prefix other than in milliliter, mL)
time	second	s	minute (min), hour (h), day (d), week, and year
Periodic and Related Phenomena			
frequency	hertz	Hz	revolutions per second (r/s)
rotational frequency	revolutions per second	s^{-1}	revolutions per minute (r/m)
Mechanics			
mass	kilogram	kg	
density	kilogram per cubic meter	kg/m^3	
moment of inertia	kilogram · $meter^2$	$kg \cdot m^2$	
force	newton	N	
moment of force (torque)	newton-meter	$N \cdot m$	
pressure and stress	pascal	Pa	(pascal = newton per square meter)
energy, work	joule	J	kilowatt-hour ($kW \cdot h$)
power	watt	W	
impact strength	joule	J	
section modulus	$meter^3$	m^3	
moment of section (second moment of area)	$meter^4$	m^4	
fracture toughness	$Pa \cdot \sqrt{m}$		

Heat			
temperature—thermodynamic [Note (2)]	kelvin	K	degree Celsius (°C)
temperature—other than thermodynamic	degree Celsius	°C	kelvin (K)
linear expansion coefficient	meter per meter-kelvin	K^{-1}	$°C^{-1}$
quantity of heat	joule	J	
heat flow rate	watt	W	
thermal conductivity	watt per meter-kelvin	W/(m · K)	W/(m · °C)
thermal diffusivity	square meter per second	m^2/s	
specific heat capacity	joule per kilogram-kelvin	J/(kg · K)	J/(kg · °C)
Electricity and Magnetism			
electric current	ampere	A	
electric potential	volt	V	
current density	ampere per meter^2	A/m^2	
magnetic field strength	ampere per meter	A/m	

NOTES:

(1) Conversion factors between SI units and U.S. customary are given in SI-1, "ASME Orientation and Guide for Use of SI (Metric) Units," and ASTM E 380.

(2) Preferred use for temperature and temperature interval is degrees Celsius (°C), except for thermodynamic and cryogenic work where kelvins may be more suitable. For temperature interval, 1 K = 1°C exactly.

Reprinted by permission of ASME.

Commonly Used Conversion Factors

(For Others See ASTM E 380) (See Note 1)

Quantity	To convert from	To	Multiply by [Note (2)]	
plane angle	degree	rad	1.745 329	E–02
length	in	m	2.54*	E–02
	ft	m	3.048*	E–01
	yd	m	9.144*	E–01
area	in^2	m^2	6.451 6*	E–04
	ft^2	m^2	9.290 304*	E–02
	yd^2	m^2	8.361 274	E–01
volume	in^3	m^3	1.638 706	E–05
	ft^3	m^3	2.831 685	E–02
	US gallon	m^3	3.785 412	E–03
	Imperial gallon	m^3	4.546 09	E–03
	liter	m^3	1.0*	E–03
mass	lbm	kg	4.535 924	E–01
	ton (metric) (mass)	kg	1.000 00*	E+03
	ton (short 2000 lbm)	kg	9.071 847	E+02
force	kgf	N	9.806 65*	E+00
	lbf	N	4.448 222	E+00
bending, torque	kgf · m	N · m	9.806 65*	E+00
	lbf · in	N · m	1.129 848	E–01
	lbf · ft	N · m	1.355 818	E+00
pressure, stress	kgf/m^2	Pa	9.806 65*	E+00
	lbf/ft^2	Pa	4.788 026	E+01
	lbf/in^2 (psi)	Pa	6.894 757	E+03
	$kips/in^2$	Pa	6.894 757	E+06
	bar	Pa	1.0*	E+05
energy, work	Btu (IT) [Note (3)]	J	1.055 056	E+03
	ft · lbf	J	1.355 818	E+00
power	hp (550 ft · lbf/s)	W	7.456 999	E+02
fracture toughness	ksi $\sqrt{in.}$	Pa · $\sqrt{m}$	1.098 843	E+06
temperature	°C	K	$t_K = t_C + 273.15$	
	°F	K	$t_K = (t_F + 459.67)/1.8$	
	°F	°C	$t_C = (t_F - 32)/1.8$	
temperature interval	°C	K	1.0*	E+00
	°F	K or °C	5.555 555	E–01

NOTES:

(1) Care should be taken when converting formulas or equations that contain constant terms or factors. The value of these terms must be understood and may also require conversion.

(2)(a) Relationships that are exact in terms of the base units are followed by a single asterisk.

(b) The factors are written as a number greater than 1 and less than 10 with 6 or less decimal places. The number is followed by the letter E (for exponent), a plus or minus symbol, and two digits which indicate the power of 10 by which the number must be multiplied to obtain the correct value. For example: 3.523 907 E–02 is 3.523 907 × 10^{-2} or 0.035 239 07.

(3) International Table.

Reprinted by permission of ASME.

Appendix

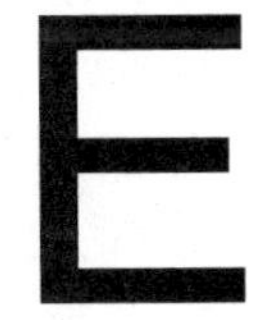

Study Guide for National Board and A Endorsement Examinations

This study guide is intended to assist you in preparing for the National Board examination to obtain a boiler inspector's commission, and the A endorsement examination to obtain the A endorsement to the National Board commission, which is required for inspecting new work. The National Board examination typically consists of one full eight-hour day of open-book (computational) questions, followed the next day by a four-hour closed-book (memory) segment. The open-book portion is ten questions, counting for 40 percent of the exam grade. The closed book portion is 30 questions, counting for 60 percent of the examination.

The ten open-book computational questions will come from Section I and Section VIII (Division 1). Roughly half will come from each section. Most of the closed book questions are taken from Section I (Power Boilers) of the ASME Code, or from the National Board Inspection Code. There are, however, a few from Sections IV, V, VIII, IX, and B31.1.

The objective of the first portion of this study guide is to structure your reading of the ASME Code and the NBIC. The format will be a reading assignment, followed by a group of questions, the answers of which may be found in the reading.

Additionally, the A endorsement examination may be given in the afternoon of the second day of the National Board exam. The format for the A endorsement examination is multiple choice, again with the questions coming from Sections I, IV, V, VIII Division 1, and IX. Additional reading and questions are given herein on Sections IV, V, and IX to assist in preparing for the A endorsement examination.

Section I

1995 Edition

Read the Foreword, the Statements of Policy, and the Preamble to then answer the following:

1. What is a power boiler?
2. What is an electric boiler?
3. What is a high-temperature water boiler?
4. What are the design rules for an unfired steam boiler?
5. What requirements must be met for the expansion tank of a high-temperature water boiler?
6. What is boiler external piping?
7. Is Code certification required for the boiler external piping?
8. What is the ASME policy on the use of Code symbols in advertising?
9. What is the ASME policy on the use of ASME marking to identify manufactured items?
10. Who formulates the rules of the ASME Code?
11. What is the objective of the rules?
12. When are revisions to the Code published?
13. When may addenda be used?
14. When do addenda become mandatory?
15. What is the National Board of Boiler and Pressure Vessel Inspectors?

Read PG-1 through PG-29 and answer the following:

16. What Code section applies to steam boilers at or below 15 psig operating pressure?
17. Is it permissible to construct and stamp such boilers in accordance with Section I?
18. What Code section covers hot water boilers not exceeding 160 psi and/or 250°F?
19. Where are the stress values for materials found?
20. What are the service limitations for gray cast iron?
21. What are the service limitations for cast nodular iron?

22. What are some allowable specifications for nipples (short pieces of pipe)?
23. What are the limitations on the use of austenitic stainless steel in boilers?
24. Can material not fully identified be used under Section I?
25. Give some examples of some nonwelded pressure parts for which inspection, mill test reports, and partial data sheets are not required.
26. Can austenitic stainless steel be used for a gage glass body?
27. What specifications are permitted for stays?
28. What is the minimum thickness of a boiler plate?
29. What is the minimum thickness of a plate to be stayed?
30. What is the minimum thickness of pipe to be used as a shell?
31. What is the permissible undertolerance on plate?
32. What is the permissible undertolerance on tubes and pipes?
33. When may a hydrostatic test be used to determine the maximum working pressure?
34. Under what conditions may a boiler exceed the MAWP? By how much?
35. When must stresses due to hydrostatic head be taken into account?
36. What quality factor applies to a steel casting inspected in accordance with the minimum requirements of the material specification?
37. What NDT methods are used on castings?
38. May the casting be repaired by welding?
39. What certification is required of the NDT personnel?
40. What added thickness is required for tubes attached by threading?
41. What is the minimum thickness of a tube for installation of a fusible plug? Why?
42. What is the minimum design temperature for a tube that absorbs heat? That does not absorb heat?
43. What weight of pipe is required for threaded applications in steam service over 250 psi?
44. What is Note 10, and when does it apply?

45. What thickness is added to a dished head with a flanged-in opening above __ inch(es) diameter?

Read PG-30 to PG-59 and answer the following questions:

46. What is the minimum drain connection size for a high-temperature water boiler?
47. How shall the boiler setting opening be arranged for a blowoff pipe?
48. What are the maximum and minimum sizes for a blowoff line?
49. For the boiler external piping, the "materials, _____, _____, _____, and _____ shall be in accordance with _____."
50. What is the minimum size of a tell-tale hole in a staybolt?
51. What is the maximum pitch for staybolts?
52. What is the minimum thickness of a plate to be stayed?
53. What is the minimum size of a manhole opening?
54. What is the minimum size of a handhole opening?
55. What is the largest diameter that may be attached by expanding?
56. What is the maximum temperature for the use of threaded connections on a boiler?
57. What is the maximum pressure for the use of threaded connections on a boiler?
58. What is the minimum distance between centers of two adjacent openings not in a definite pattern?
59. What is the area required for compensation for an opening in a shell? In a flat head?
60. What are the limits of the rules for compensation given in Section 1?
61. May hub-type flanges be machined from plate material?
62. What is the minimum thickness of a nozzle neck used in a compensated opening?
63. What is the requirement for the ligament between tube holes around the circumference of a drum?

Read PG-60 to PG-113 and answer the following questions:

64. When are two gage glasses required on a power boiler?
65. Under what conditions may the gage glass on a boiler be shut off?
66. What is the minimum size of pipes to a water column?

67. What are the maximum pressures for cast iron water columns? For ductile iron water columns?
68. What are the material specifications for gray iron water columns? For ductile iron water columns?
69. May shutoff valves be used in the pipe connections between a boiler and a water column? (If so, what are the requirements?)
70. Are gage cocks or try cocks required? (If so, how many?)
71. What should be the range of the pressure gage?
72. What is the difference between the pressure gage requirements for a steam boiler and a high-temperature water boiler?
73. What valves are required and what valves are permitted in the gage connection to the boiler?
74. What gage is required on a hot water boiler that is not required on a steam boiler?
75. What size is required for the test valve connection?
76. Under what conditions does a boiler require more than one safety valve?
77. What safety valve capacity is required?
78. If there are two safety valves, what is the highest pressure at which one valve can be set? What is the lowest pressure at which the other valve can be set?
79. What type of safety valves are required on steam boilers?
80. Can dead-weighted safety valves (also called weight-and-lever safety valves) be used? If so, under what conditions?
81. Can safety valves be attached by threading? If so, what are the limitations?
82. What type of safety valve is required on a high temperature water boiler?
83. Can the superheater safety valve be counted as part of the boiler relieving capacity? If so, under what limits?
84. What is the minimum relieving capacity for an electric boiler?
85. Where is heating surface measured?
86. What are the three ways of measuring (i.e., testing) the required safety valve capacity of a boiler?
87. Can a muffler be used on a safety valve?
88. How much may the spring of a safety valve be reset?
89. After blowing down, the safety valves should close at what pressure?

90. May the end faces of nozzles be cut by shearing?
91. Under what circumstances may the vessel manufacturer transfer the markings on a plate?
92. Under what circumstances may a representative of the plate manufacturer transfer the markings on a plate?
93. What must be done if required heat treatment is not performed by the plate manufacturer (i.e., the mill)?
94. What is required if defective material is to be repaired?
95. What is required in the making of tube holes in plate?
96. What is the maximum permissible out-of-roundness for a drum or shell? For a circular furnace? Why the difference?
97. What are the requirements for holes for screwed stays?
98. Who is responsible for making the design calculations?
99. What are the qualifications of the Authorized Inspector?
100. What is the maximum stress on any part of a boiler during the hydrostatic test?
101. What is the nominal pressure for the hydro? What pressure must not be exceeded in the hydrostatic test?
102. What are the minimum and maximum water/metal temperatures during the hydrostatic test? What are the reasons for these temperature requirements?
103. When may a hydrostatic test be used to determine the MAWP?
104. What are the six Code stamps?
105. How does a company apply for a Certificate of Authorization?
106. Do safety valve manufacturers need a contract with an Authorized Inspection Agency?
107. Where are the requirements for the quality assurance system found? (Write out these requirements—it is often a closed book question.)
108. Who conducts the review of an applicant for a certificate of authorization?
109. Can a manufacturer start fabricating code vessels before receiving a certificate of authorization? If yes, what are the limitations?
110. Who must be present at the stamping of boilers?
111. When may a boiler not be stamped?
112. What information must be on the nameplate or stamping in addition to the Code symbol?

113. Who can fabricate the boiler external piping?
114. Who can install the boiler external piping?
115. Where should the stamping be on a Scotch marine boiler? On an Economic boiler? On an HRT boiler? On a watertube boiler?
116. Who is responsible for distributing copies of the Master Data Report Form?

Read Part PW and answer the following questions:

117. Can a manufacturer use welders not employed by the manufacturer? If so, what are the conditions?
118. What is the maximum carbon content for steels to be welded?
119. When butt welding materials of different thickness, must the centerlines of the two pieces coincide? Explain.
120. How may a single-welded butt weld be made equivalent to a double-welded butt weld?
121. Under what conditions may ultrasonic examination of a weld be substituted for radiography?
122. What flange lengths are required for heads over 24 inches? Under 24 inches?
123. Is it permissible for a reinforced opening to be placed through a welded joint?
124. May nozzles be attached by gas welding?
125. What is the size and purpose of the tell-tale hole in a reinforcement pad?
126. What factor is applied to the stress value from Section II, Part D for nozzle compensation for groove weld shear?
127. What is the minimum fillet weld size for a diagonal stay?
128. Are tell-tale holes required in welded staybolts?
129. When may radiography of a production weld be used to qualify a welding operator?
130. What records must the manufacturer maintain with regard to the welding?
131. Do tack welds require a qualified procedure? Qualified operators?
132. What is the maximum permitted undercut?
133. Can backing strips be left in place?
134. When do welded parts require postweld heat treatment?
135. What are the time and temperature requirements for PWHT on P-1 materials?

136. When is radiography of butt welds not required in the boiler proper?
137. It is the duty of _____ to assure himself that the welding procedures have been qualified.
138. When is a tube not subject to radiant heat?
139. When is a tube not subject to furnace gases?
140. What are the acceptance standards for radiographic examination of welds?
141. What are the acceptance standards for ultrasonic examination of welds?
142. To what edition of SNT-TC-1A shall the NDT personnel be qualified?
143. When are vessel plates required to be made?
144. What tests are required on the vessel plates?
145. Can nonpressure parts be welded to pressure parts after the hydrostatic test?
146. Can riveted boilers be built in accordance with ASME Section I? If so, to what edition?

Read Part PWT and answer the following questions:

147. What materials or material specification is mandatory for mud drums of watertube boilers?
148. When can table PWT-10 be used?
149. How far should the tubes project into the drum?
150. Can threaded pipe be used as boiler tubes?
151. What types of latches are not permitted on firing doors of watertube boilers if the doors open outward?
152. What is the minimum size of an access door in the boiler setting?

Read Part PFT and answer the following questions:

153. What is the minimum thickness of the shell of an 84-inch HRT?
154. What is the minimum thickness of the tubesheet for the 84-inch HRT?
155. What are primary furnace gases?
156. Can the rear head of an HRT boiler be unflanged and attached to the shell by welding?
157. Sketch and dimension an Adamson ring.

158. What is the minimum section length of an Adamson furnace?
159. What is the minimum thickness of the Adamson furnace?
160. How is the thickness of a corrugated furnace gaged?
161. What is the area supported by a stay?
162. If a flanged-in manhole is placed in the flat stayed head of a firetube boiler, what is the reduction in the area to be stayed?
163. If sling stays are used with crown bars, what proportion of the load shall the stays be designed to carry?
164. What is the minimum size of washout plugs?
165. What is the minimum size of access doors into the boiler setting?
166. What inspection openings are required on a Scotch boiler?
167. May a valve be placed between the safety valve and the boiler to facilitate repairs to the safety valve? (If so, what are the limitations?)
168. An HRT boiler over _____ diameter must be supported by steel hangers, independent of the furnace side walls.
169. On wetbottom boilers, what is the minimum distance between the shell and the floor? Why?
170. When the water in the gage glass is at its lowest reading, how much water is over the tubes of a horizontal firetube boiler?
171. If the blowoff piping is exposed to the products of combustion, how shall it be attached to the boiler?
172. If the blowoff piping is not exposed to the combustion products, how may it be attached?

Read parts PFH, PMB, PEB, and PVG and answer the following questions:

173. How may the feedwater heater be constructed?
174. What is a miniature boiler?
175. What is the minimum thickness of heads and shells for miniature boilers?
176. Are radiography and postweld heat treatment required for miniature boilers?
177. What is the hydrostatic test pressure for miniature boilers?
178. What miniature boilers do not require a low-water fuel cutoff?
179. What methods of construction may not be employed by an E stampholder?

180. What is the minimum thickness for boiler plate and heads in electric boilers?
181. In what type of Section I vessel are lifting levers on the safety valves prohibited?
182. What is the minimum size of the blowoff connection on a miniature boiler?
183. What steam boilers do not require washout or inspection openings?
184. The design pressure of an organic vapor generator shall be at least ___ psi above the operating pressure?

Read A-19 to A-64 and answer the following questions:

185. How often must fusible plugs be replaced?
186. How thick must a tube be if a fusible plug is to be inserted?
187. Where are fusible plugs (if used) required to be located?
188. When are proof tests used to establish the MAWP?
189. What are the two types of proof tests?
190. When a proof test has established the MAWP of a part, what pressure testing is required on duplicate parts?
191. Who shall witness the proof test?
192. What is the maximum range of set pressures of the safety valves on a steam drum?
193. How may the required safety valve capacity be checked?
194. What valves may be used between the safety valve and the boiler?

National Board Inspection Code (NBIC)

1995 Edition

The 1995 edition of the National Board Inspection Code represents a complete reorganization of this code book. In addition to the renumbering and rearranging, there have been a great number of substantive changes as well.

Read pages viii through xiii and Part RA and answer the following questions:

1. What is the National Board?
2. When are addenda to the NBIC published?

3. What is the purpose of the NBIC?
4. What are the certificates of authorization issued by the National Board?
5. What are the six types of repair authorization?
6. What are the prerequisites for obtaining an R stamp?
7. Where does a repair organization actually obtain an R stamp?
8. How often must an R stamp be renewed?
9. What is the composition of an R stamp review team?
10. What are the prerequisites for an owner-user organization?

Read Appendix 4 (Glossary of Terms) and answer the following questions:

11. What is a repair?
12. What is an alteration?
13. What is a jurisdiction?
14. What is an Authorized Inspection Agency?
15. What is an owner-user inspection organization?

Read Part RB and answer the following questions:

16. How should the owner prepare a boiler for internal inspection?
17. What light is preferred for internal inspection?
18. What other light may be used?
19. If a small amount of oil is found on the waterside surfaces, is the condition serious? What, if anything, should be done?
20. How should staybolts be examined and tested?
21. What is a bulge?
22. What is a blister?
23. May a bulge on a plate be driven back?
24. Should fire cracks be repaired?
25. What cracks may not be repaired?
26. What are the most common causes of corrosion in boilers?
27. In vertical firetube boilers, where is excessive corrosion and pitting found?
28. How should the pipes to the water column be pitched for drainage?

29. Where should the blowoff pipe discharge?
30. For what should the Inspector check in examining the fireside of watertubes?
31. How often should safety valves set at under 400 psi be operationally tested?
32. What types of hydrostatic test may be conducted?
33. What are the maximum and minimum water temperatures for the hydrostatic test?
34. What pressure vessels may require only an external inspection?
35. What are the five forms of corrosion?
36. When is an oxygen test required before entering a pressure vessel?
37. What is the proper range of Oxygen readings?
38. What is the range of hydrostatic test temperatures for pressure vessels?
39. How are the safety valves prevented from lifting during a hydrostatic test at 1.5 times the MAWP?
40. How should a low water fuel cutoff device be examined?

Read Parts RC and RD and Appendix 2 and 6 to answer the following questions:

41. Who must authorize repairs?
42. Who should make the acceptance inspection of the repair or alteration?
43. What NDE is required for welded repairs and alterations?
44. What is the maximum carbon content of steels to be welded?
45. Is a pressure test required after the completion of a repair?
46. Is a pressure test required after an alteration?
47. What is done if defective bolting is found?
48. Can wasted areas around manhole openings be built up by welding?
49. Is seal welding of riveted joints allowed?
50. Sketch a flush patch.
51. Sketch a window patch.
52. Who prepares form R-1?
53. Is form R-1 required for all repairs?

54. Where is the required repair stamping applied to the boiler or pressure vessel? By whom?
55. Who signs form R-2 for an alteration?
56. Who may perform an alteration?
57. What is rerating?
58. Who performs the calculations for a rerating?
59. Sketch the stamping applied to a rerated boiler.
60. What is the size of the stamping?
61. What organizations may perform alterations?
62. Give three examples of repairs of a routine nature in a boiler.
63. Give three examples of alterations to boilers.
64. What methods may be used as alternatives to postweld heat treatment?
65. What repairs must be reviewed and certified by a professional engineer?
66. Under what conditions may a repair company use a welding procedure that it has not qualified?
67. When may the Inspector require requalification of a welder?
68. What should the AI look for in checking completed welds?
69. What qualification is required for NDE personnel?
70. What should the Inspector check in examining piping?
71. How should the gage glass be tested?
72. Who must approve the procedures, qualifications, and organization of an owner-user inspection agency?
73. What is the maximum period between inspections for pressure vessels?
74. What is the formula for remaining life of a pressure vessel?

Read Appendix F and answer the following questions:

75. Why must there be a difference between the safety valve set pressure and the normal operating pressure?
76. What should the differential be for a hot water heating boiler? For a steam heating boiler?
77. What, then, is the normal maximum operating pressure for most steam heating boilers?
78. If a boiler has a MAWP of 400 psi, what should be its maximum operating pressure?

79. What is the recommendation for the MAWP and safety relief valve setting for high temperature hot water boilers?
80. What is the recommended safety valve testing frequency for power boilers up to 400 psi MAWP?
81. Who files the boiler or pressure vessel identification form?
82. What is the purpose of the form?
83. What is the subject of Appendix 3?
84. From what code is Table L-2 taken?
85. Who owns the R stamp?
86. What special requirements apply to repair of pressure vessels constructed to Section VIII, Division 2?
87. Who must approve restamping of a boiler?
88. If the nameplate is lost or illegible, what form must be used in applying to restamp?
89. Who must restamp the Code symbol and National Board number?
90. Who must witness such stamping?
91. What is the minimum safety factor for noncode boilers?
92. What are the accepted routine repairs?
93. When is installation of a new nozzle a repair?
94. What is the most common failure mechanism in deaerator heaters?
95. What NDE method is recommended for deaerator heaters?
96. What type of cracking are liquid ammonia vessels especially susceptible to?
97. What NDE method is recommended for what areas of liquid ammonia vessels?
98. What is the preferred method of repair if cracks are found in liquid ammonia vessels?
99. What should one look for in checking piping supports?
100. How does one determine the thickness of piping system components subject to erosion or corrosion?
101. With what code must the locking devices of quick-actuating closures comply?
102. What warning devices are required on quick-opening doors with manually operated locking devices?

103. What is the special hazard associated with quick-opening closures?
104. What items should be checked in inspecting quick-opening closures?

Section IV: Heating Boilers

1995 Edition

Read the Foreword and the Statements of Policy and answer the following questions:

1. What is the function of the Boiler & Pressure Vessel Committee?
2. What is the objective of the Code rules?
3. How must technical inquiries to the Boiler & Pressure Vessel Committee be prepared?
4. When do the addenda become mandatory?
5. What is the membership of the National Board of Boiler and Pressure Vessel Inspectors?
6. Can the ASME Code symbol be used in advertising? If so, under what conditions?

Read the Contents, the Preamble, and Articles 1 and 2 to answer the following questions:

7. What kinds of boilers are covered by Section IV?
8. What are the service restrictions for Section IV boilers?
9. Can materials not identified by a mill test report be used in Section IV boilers? If yes, what are the limitations?

Read pages HG-300 to HG-328.2 and answer the following questions:

10. What is the design pressure per Section IV? What is its minimum value?
11. What is the relationship between the design pressure and the MAWP?
12. What are the three types of formed heads in Section IV?
13. May pipes be used as tubes?
14. What is the minimum knuckle radius of a torispherical head?
15. What is the maximum crown radius of a torispherical head?
16. What is the maximum pressure on the convex side of a dished head?

17. What is the minimum thickness of a cylindrical furnace?
18. What is the maximum deviation from a true circle for a rolled circular furnace?
19. What is the size of flanged opening above which a thickness increase of ⅛ inch or 15 percent is required?
20. What is the minimum gasket seating width for a flanged-in manhole?
21. What is the minimum flange depth for a flanged-in opening? How is the depth of flange measured?

Read pages HG-328.3 to HG-360 and answer the following questions:

22. What boilers do not require washout plugs or inspection openings?
23. What is the minimum size of a manhole?
24. What is the maximum gasket thickness for a manhole or handhole?
25. What is the minimum size of a handhole?
26. What length of staybolt requires what size tell-tale hole?
27. Are tell-tale holes required in welded staybolts?
28. Can stays be fabricated by welding?
29. How are tube holes to be made?
30. What is the minimum number of pipe threads on a curved surface?

Read pages HG-360.2 to HG-402.6 and answer the following questions:

31. What stamping may be on the safety valve of a Section IV boiler?
32. What is the maximum pressure rise above the MAWP when the safety valves are operating on a steam boiler?
33. Where is heating surface measured?

Read pages HG-500 to HG-540.8 and answer the following questions:

34. When is proof testing used to establish the design pressure?
35. What are the types of proof testing?
36. What is the proof test for parts subject to collapse?
37. Who must witness the proof test(s)?
38. What is the hydrostatic test pressure for all but cast iron boilers?
39. What are the responsibilities of the AI with respect to H boilers?
40. Who must approve changes to a Manufacturer's quality assurance system?

Read pages HG-600 to HG-715 and answer the following questions:

41. What gage is required on a hot water boiler that is not needed on a steam boiler?
42. What instruments must be visible at all times?
43. May a shutoff valve be installed between the safety valve and the boiler? If so, what are the requirements?
44. In hot water systems, is there a requirement for an expansion tank?
45. What capacity is required for an expansion tank? What design pressure? What design code?
46. What capacity and design code are required for the expansion tank in a hot water supply system?

Read pages HG-716 to HF-302 and answer the following questions:

47. Can 300-series stainless steel be used in Section IV boilers? If so, what are the limits?
48. What is the minimum thickness of ferrous plate or pipe used in lieu of plate?
49. Where are the maximum allowable stress values for Section IV construction found?
50. What is the basic safety factor for Section IV construction?

Read pages HW-400 to HW-731 and answer the following questions:

51. What materials are permissible for welded construction?
52. What is the maximum carbon content of material to be welded?
53. How shall procedures, welders, and welding operators be qualified?
54. Can production welding be done prior to qualification?
55. When welding plates of unequal thickness, what are the requirements? (Sketch.)

Read HC-300 to HC-510 and answer the following questions:

56. What are the hydrostatic test requirements for cast iron steam boilers?
57. What are the hydrostatic test requirements for cast iron hot water boilers?

Read the Introduction to Part HLW and answer the following questions:

58. What is a Part HLW boiler?
59. What are the maximum temperature and pressure?
60. What is the maximum volume? (See HLW-100 to HLW-102.)

Section V: Nondestructive Testing

1995 Edition

Read T-110 to T-190 and Appendix 1 and answer the following questions:

1. How shall NDT personnel be qualified?
2. How shall NDT personnel be qualified if Section V, Article 1 is referenced by the Code section?
3. How may NDT personnel be qualified in the absence of specific qualification requirements and in the absence of a reference to Section V, Article 1?
4. What edition of SNT-TC-1A is used per Section V?
5. When must all NDT examinations be performed to written procedures?
6. Who is responsible for providing the NDT procedures and personnel certifications required?
7. Who performs inspections?
8. Who performs examinations?
9. Where are acceptance standards found?
10. What records/documentation is required?
11. Who is responsible for the records/documentation?
12. What NDT methods are within the scope of Section V?

Read T-210 to T-292 and answer the following questions:

13. When Article 2 is required by the referencing Code section, how may compliance be demonstrated?
14. What surface preparation is required for radiography?
15. Why is a lead letter B required on the back of film holders?
16. What is excessive backscatter (or how is it determined that backscatter is excessive)?
17. How is film density measured?

18. What film must be used for Section V radiography?
19. What types of IQI may be used?
20. How are densitometers calibrated?
21. When shall double-wall radiography be used?
22. What is the minimum thickness of steel to be radiographed with IR-192?
23. What is the formula for geometric unsharpness? What does each term in the formula stand for? What are the units of geometric unsharpness?
24. Where shall location markers be placed?
25. Where shall the IQI be placed?
26. When the IQI cannot be placed as in question 25, how shall it be placed and identified?
27. What is a shim made of, and what is it for?
28. If a circumferential seam is radiographed using central source placement and exposing all films at once, how many penetrameters are needed?
29. What is the minimum density for a radiograph made with x-ray? With an isotope?
30. What is the maximum density?
31. What is the minimum density for each film in composite viewing?
32. What is the maximum permissible density range of a film? What must be done if this range is exceeded?
33. Is a dark image of a B on a lighter background reason for rejection of a film?
34. What is the maximum permissible geometric unsharpness for material under 2 inches thickness per Section V? Per Section I?

Read T-410 and T-510 to T-594 and answer the following questions:

35. What is the subject matter of Article 4? Of Article 5?
36. Is a written procedure required for UT inspection of materials and fabrications?
37. How are forgings and bars inspected ultrasonically?
38. What is a DAC curve?
39. What frequency is (nominally) used?
40. What is the hole size for the basic calibration block for ultrasonically inspecting welds up to 1 inch thick?

Read T-600 to T-690 and Appendix 6 and answer the following questions:

41. What are typical discontinuities found by liquid penetrant testing?
42. What is the metal temperature range for liquid penetrant inspection?
43. What are penetrant materials?
44. What six techniques of PT are used per Section V?
45. How may penetrant be applied?
46. What is the maximum water temperature and pressure for water washable PT?
47. What is black light?
48. When shall final interpretation be made in PT?
49. In the solvent removable method, how is cleaning performed after the penetration time is up?
50. What is the minimum black light intensity on the surface of the part in fluorescent PT?

Read T-710 to T-791 and Article 7, Appendix I and II and answer the following questions:

51. What does the magnetic particle examination method detect?
52. What method of examination is required per the Code?
53. What five techniques may be used?
54. How shall magnetic flux be produced in the part?
55. What is the maximum temperature for dry powder MT?
56. What is the maximum temperature for wet method MT?
57. How long must the examiner be in the darkened area before interpreting results in fluorescent MT?
58. How long must the black light be warmed up before use?
59. What must be the lifting power of an AC yoke? Of a DC yoke?

Section IX: Welding and Brazing Qualifications

1995 Edition

Read the Introduction and QW-100 to QW-196.2.2 and answer the following questions:

1. What is the purpose of the WPS and PQR?
2. What is in the WPS?

3. What is in the PQR?
4. What is a welder?
5. What is a welding operator?
6. All WPSs, PQRs, and performance qualifications made in accordance with the _____ edition of Section IX may be used in any construction per the ASME Boiler & Pressure Vessel Code and B31.1.
7. Who is responsible for the welding, and who must conduct the tests?
8. Who must maintain the records of the welding tests?
9. When may radiographic examination be used for qualification?
10. If the tensile test sample breaks in the parent metal, what is the minimum acceptable strength?
11. When may longitudinal bend tests be used?
12. What is the maximum opening permitted in the bend test samples?
13. What testing is required for fillet weld procedure qualification?
14. What is the maximum difference in the legs of the fillet weld?
15. What is the maximum length of a slag inclusion in a welder performance test radiograph?

Read QW-200 to QW-251.4 and answer the following questions:

16. Where does one find the variables that must be listed in the WPS?
17. Where must the WPS be available?
18. Who certifies the accuracy of the PQR?
19. Where are the limits of qualification found?
20. A qualification in any position qualifies the procedure for all positions. True or false?

Read QW-300 to QW-384 and answer the following questions:

21. How may a welder or welding operator be qualified?
22. Can a welder qualify for more than one organization at a time?
23. What is the intent of the performance qualification test?
24. Is preheat or postheat treatment required in a performance test if such treatment is required by the WPS?
25. What is the minimum length of weld to be examined for a qualification test by radiography?

26. How many bend test specimens are required for a pipe performance qualification in the 1G or 2G positions? In the 5G or 6G positions?
27. Does a groove weld qualification also qualify for fillet welding?
28. Is a welder qualified to weld per one WPS also qualified to weld in accordance with other qualified WPSs using the same process? If so, what are the limits?
29. When must a welder be requalified (per QW-351)?
30. When does a welder lose his qualification?

Read QW-400 to QW-401.5 and answer the following questions:

31. What is an essential variable for procedure? For performance?
32. What is a supplemental essential variable?
33. What is a nonessential variable?
34. Where does one find the variables listed?
35. Where does one find whether a variable is essential or nonessential?

Read QW-403.1, QW-420.1, QW-430, and QW-431 and answer the following questions:

36. How are base metals grouped?
37. How are electrodes grouped?
38. What are the F numbers for steel and steel alloys? (See page 120.)

Read QW-450, QW-451, and QW-452 and answer the following questions:

39. When may side bends be substituted for required face and root bends in a procedure qualification?
40. How many tensile tests are required in a groove weld procedure qualification?
41. If the test coupon is ⅜ inch thick, what is the range of thickness qualified?
42. If the test coupon is 2½-inch NPS pipe, what diameter range is qualified?

Read QW-460 and QW-461 and answer the following questions:

43. Sketch and label the test positions for groove welds in plate.
44. Sketch and label the test positions for groove welds in pipe.

45. Sketch and label the test positions for fillet welds in plate.
46. Does welding pipe qualify a welder to weld plate?
47. Does welding plate qualify a welder to weld pipe? If so, what are the limits?
48. What test plate positions are required to qualify to weld all positions?
49. What two ways may a welder qualify to weld all positions in pipe?
50. Definitions are found in what paragraph?
51. Where may the forms for the WPS, PQR, and WPQ be obtained?

B31.1: Power Piping

1995 Edition

Read 101.1 to 101.5.1, 102.3.1, and 102.3.2 A and B and answer the following questions:

1. In designing boiler external piping, does static head have to be considered?
2. What is the value of the metal temperature?
3. What is water hammer?
4. Where are the allowable stresses for boiler external piping found?

Read 103 and 104.1 and answer the following questions:

5. What does the term A in the formulae of 104.1 represent?
6. How is pipe thickness calculated for use under external pressure?
7. Above what pressure for steam service must pipe be seamless and at least schedule 80?

Read 122.1 to 122.6 and answer the following questions:

8. How shall blowoff piping of firetube boilers be attached if it is exposed to the products of combustion? If it is not so exposed?
9. If slip-on flanges are used, what is the maximum size? What is the minimum weld size?
10. What is the design pressure for the piping from the superheater header to the first stop valve? The temperature?
11. What is the lowest permissible design temperature for any steam piping in the boiler external piping?

12. What pressure is required for the design of the feedwater piping from the boiler to the required check valve?
13. What is the design temperature for the feed piping?
14. What is the minimum design pressure for the feedwater piping?
15. How are blowoff piping systems operated?
16. How are blowdown systems operated?
17. What is the design pressure and temperature for a blowoff pipe?
18. Above what pressure is schedule 80 required for the blowoff line?
19. What is the minimum design pressure for a blowdown pipe?
20. What is the required design pressure for a drain pipe?
21. What steam stop valves are required if the boiler is not connected to a common header? If it is connected to a common header?
22. What kinds of valves may not be used as blowoff valves?
23. May a combination stop and check valve be used in place of the two valves specified for the feed line?

Glossary

0.2 percent offset yield strength. That value of stress that will produce 0.2 percent permanent strain in a material. This is permanent strain of 0.002 in/in.

A endorsement. Endorsement or addition to the National Board Commission required for AI's performing inspection of new construction.

accumulation test. Test to verify that safety valve capacity is adequate. Boiler is fired at maximum rate with all steam outlets closed. The MAWP is not to be exceeded by more than 6 percent. See A-46.1.

Adamson ring. Type of ring reinforcement for furnaces under external pressure. See PFT-16.

AI. Authorized Inspector.

AIA. Authorized Inspection Agency.

appurtenances. Devices commonly fitted to boilers such as valves, pressure gages, and water columns.

ASME designee. Person designated by the ASME to review safety valve manufacturer's facilities, designs, and documents, or to participate in joint reviews of Manufacturers.

ASNT. American Society for Nondestructive Testing.

assembler. Holder of the A assembly stamp. This individual assembles but does not manufacture boilers and may install boiler external or other piping.

ASTM. American Society for Testing & Materials.

authorized inspection agency. A Jurisdiction having the Code as a legal requirement or an insurance company licensed to insure boilers and pressure vessels. See NBIC.

authorized inspector. Employed as an Inspector by an Authorized Inspection Agency and qualified by written examination under the rules of any state or province. See PG-91 and the NBIC.

authorized observer. Person in charge of safety valve testing, subject to ASME approval.

B endorsement. Endorsement to the National Board Commission for supervisors of AIs. An AI doing A endorsement work must report to a supervisor holding the B endorsement.

backing strip/ring. Strip or ring of steel placed along the root of a weld to facilitate making a full-penetration butt weld from one side.

baffle. Refractory structure in a boiler furnace used to control the flow of combustion gases.

bend tests. Required for the vessel plates per PW-53.9. Sample is bent cold and must not crack or contain more than a 1/16-inch opening.

blank head. Flat or formed head having no opening(s).

blind flange. Flat cover having no opening.

blister. Deformation of part of the thickness of a boiler plate or tube caused by local overheating of an area containing a lamination.

blowdown line. Pipe and valve arrangement for continuously removing water from the boiler to reduce the concentrations of dissolved solids.

blowoff line. Pipe and valve arrangement to permit periodic blowing down of the boiler to remove sediment.

blowdown. Difference between set pressure (popping point) and closing pressure on a safety valve.

boiler. Device for making steam under pressure for use external to itself.

boiler & pressure vessel committee. Committee of the ASME that writes the ASME Code. See the foreword to Section I.

boiler bushing. Threaded fitting for passing piping through a shell. See Figure PG-59.1.

boiler external piping. Piping running from the boiler proper to and including all required valves. See the Preamble and PG-58.

bourdon tube. Curved closed tube in a pressure gage. As the tube straightens, the gage hand moves.

brinell hardness. Penetration hardness measured as the area of an impression made by pressing a hardened steel ball into a surface under a known load. Typically, a 10-mm ball and a 3000-kg load are used.

Btu. British thermal unit: that amount of heat that will raise the temperature of 1 lb of water 1 degree F, 0.252 kilocalories.

bulge. Deformation of boiler plate or tube caused by overheating and the action of the pressure that affects the full thickness of the material.

cast iron. Gray cast iron meeting specification SA-278, used in some Section I components.

cock. Type of valve operated by rotating a tapered plug containing a hole in a tapered seat. Opens fully in one-quarter turn.

Code case. A temporary (three-year) permission to use a material in Code construction that may be issued by the Code Committee on application per A-82.

cogeneration. The use of several power generating devices together such as diesel engines or gas turbines, the exhaust of which goes into a heat recovery boiler.

cold short. Brittle when cold or cool.

compensation. Metal located adjacent to an opening that can carry the tensile load caused by the opening.

condensate. Steam that has returned to the liquid state.

creep. Long-term permanent deformation at stress levels below the yield stress.

critical point. Temperature and pressure where the difference between liquid and vapor phases of water disappears; 706.1°F and 3226 psia.

crown bar. Part of the system for staying the top of a combustion chamber using bolts and beams. Similar to girder stays.

crown sheet. The plate forming the top of a combustion chamber.

deadweight safety valves. Safety valves where the force holding the valve closed is supplied by weights and levers. Now prohibited by Code. See PG-67.5.

deaerator. Feedwater heater designed to remove dissolved oxygen from the feedwater.

defect. Discontinuity in a weld or material of such size that it is judged detrimental to the serviceability of the component. Rejectable discontinuity.

design pressure. Numerical value of pressure used to calculate minimum required thickness of boiler parts. Same as MAWP in Section I.

diagonal stay. Stay or brace running from a point on the circumference of the shell to support a flat head. See PFT-32.

discontinuity. Break in the normal structure of the part or material.

drum. Cylindrical pressure vessel, particularly the upper and lower drums of watertube boilers.

ductility. Ability of a material to be permanently deformed ("worked") without cracking or breaking.

E stamp. Section I stamp for manufacturing electric boilers by methods not including welding or brazing.

economizer. Type of feedwater heater absorbing waste heat from flue gases.

efficiency. Ratio of the strength of a plate or component containing a joint or holes to the strength of an identical part without the joint or holes.

elastic modulus. Also called Young's modulus. Slope of the stress-strain curve in the linear region, i.e., ratio of stress to strain.

electric boiler. Boiler in which the heat is supplied by either immersion heaters or by passing a current through the water.

ellipsoidal. See semiellipsoidal.

elongation. Elongation or percent elongation: ratio of change in length to original length in a tensile test; measure of ductility.

endurance limit. Value of stress below which a material can withstand unlimited repeated loadings. May be 35–55 percent of ultimate tensile strength.

examination. Term used to denote all forms of nondestructive evaluation performed by NDT or NDE personnel as distinguished from inspection, which is performed by the AI.

expanding. Method of installing a tube into a hole by rolling the tube end with a tool.

explosion doors. Doors built into a furnace designed to open in the event of a furnace gas explosion to vent the explosion safely and minimize damage to the boiler.

firebox steel. High-quality steel plate low in phosphorus and sulfur and used for boilers that was manufactured in the era before there were standards and specifications.

fired steam boiler. Boiler in which the source of heat is the combustion of fuel. See the preamble to the Code.

flange. Flat ring threaded or welded to the end of a length of pipe having a bolt circle for bolting up to a mating flange; *or* the cylindrical extension on a flat or formed head to facilitate attaching the head to a shell.

flange steel. High-quality steel that may be somewhat higher in sulfur and phosphorus than firebox steel.

forced-flow, no-fixed-waterline boiler. Type of once-through boiler producing steam at supercritical pressures. Example: Babcock & Wilcox "Universal Pressure" (UP) boilers.

furnace. Combustion chamber of a boiler.

fusible plug. A hollow brass or bronze threaded fitting filled with 99.3 percent pure tin intended to melt if the boiler overheats. See A-19, A-20, and Figure A-10.

gag. Device for holding a safety valve closed during a hydrostatic test; a type of clamp.

gage cocks. Set of three or more valves for testing water level in a boiler; no longer Code required. See A-51, PG-60.4.

geometric unsharpness. Measure of the inherent unsharpness of a radiograph due to source size, source to subject distance, and subject thickness. See PW-11.2.1.

girder stay. Device for supporting the crown sheet in combustion chambers by use of bolts carried by girders, the ends of which are supported on the tubesheet and backsheet.

header. A pipe or box shaped or other distribution device for supplying a number of smaller tubes.

heat-affected zone. Area adjacent to a weld, the heat treatment condition and structure of which have been altered by the heat of the welding process.

heating value. Amount of heat released in combustion of fuel per unit of fuel; e.g., Btu/lb, Btu/gal, Btu/ft^3.

hemispherical. Type of formed head, the shape of which is a complete hemisphere.

High-temperature hot water boiler. Boiler producing hot water at over 160 psig and/or 250°F.

hot short. Brittle when hot.

HRT boiler. Horizontal return tube boiler, a type of externally fired boiler having horizontal firetubes.

huddling chamber. Part of a safety valve where steam acts on an increased area after the valve first starts to open, causing the valve to pop open.

hydrostatic test. Pressure test of a boiler using water as the medium. On a new or altered boiler, the test pressure is 1.5 times the MAWP. The proof test is also a hydrostatic test.

in battery. A number of boilers with a common setting and common headers.

indication. That which is actually perceived in a nondestructive test to indicate the presence of a discontinuity.

inspection. The activities of the Authorized Inspector in his role of assuring the Code compliance of the boiler or pressure vessel. Distinguished from examination, performed by NDE personnel.

interpretations. Replies to technical inquiries to the Code committee, published in July and December.

joint review. Examination by an AI and an ASME designee of a Manufacturer's facilities, quality assurance manual, and the like to verify that a certificate of authorization should be issued or renewed.

joint tension test. Required tensile test on the vessel plate, with the sample cut from the plate such that the sample axis is transverse to the weld axis.

Jurisdiction. Governmental agency exercising legal power to administer the Code in a geographic location—see also the glossary in the NBIC; *or* the state, municipality, or province of Canada that has adopted one or more ASME Code sections as law.

killed steel. Chemically deoxidized steel.

lagging. Insulating material used to wrap pipe and other boiler components.

lamination. Two-dimensional (planar) discontinuity in steel with major dimensions parallel to the surface caused by rolling in an inclusion, porosity, or pipe.

lifting lever. Test lever for testing safety valves while a boiler is steaming.

ligament. Metal between holes drilled in a drum. See PG-52.

limits of compensation. Distance from an opening in a shell or head within which extra metal may be counted as compensation.

load factor. Value to be computed and used in calculating strength of lugs and brackets used for structural loads and welded to tubes. See PW-43.

locomotive boiler. Type of firebox boiler having a water-cooled firebox at one end and horizontal tubes running through a shell or barrel to the other end.

low-water cutoff. Also called low-water fuel cutoff. Device for shutting down a boiler in the event that the water falls below safe levels.

lug. Bracket or hanger.

MAWP. Maximum allowable working pressure.

maximum allowable working pressure. Nominal pressure not to be exceeded when a boiler is in service; same as design pressure in Section I.

miniature boiler. Boiler no larger than 16 inches in diameter, with a maximum of a 5-cubic-foot volume, a 25-square-foot heating surface, and a 100-psig MAWP. See Part PMB.

minus heads. Formed heads having pressure on the convex side.

National Board. National Board of Boiler & Pressure Vessel Inspectors; licenses boiler inspectors.

NBIC. National Board Inspection Code.

nipple. Short length of pipe or tube.

nodular iron. A form of cast ductile or nodular iron meeting SA-395. The graphite is in spherical form.

nonreturn valve. Valve allowing flow in one direction only; required in feed lines.

nonstandard pressure parts. Parts formed without welding that may be supplied essentially as materials. See PG-11.2.

nozzle. Tube or pipe inserted into an opening in a shell or head.

NPS. Nominal pipe size.

obround. Shape having semicircular ends and parallel sides. See Note 15 to PG-32.2.1.

Ogee. Type of reversed-curve flange used to join concentric cylinders together. See Figure PFT-20.

operating pressure. Normal pressure actually maintained while boiler is steaming. Must be less than MAWP.

organic fluid vaporizer. Fired pressure vessel in which organic vapor is generated by application of heat. See the Preamble.

owner-user agency. Owner-user of pressure vessels who maintains an inspection department satisfactory to both the National Board and the Jurisdiction. See NBIC.

P numbering. System of grouping parent metals of similar composition and welding characteristics.

pig tail. See syphon.

pipe. Defect in steel caused by shrinkage of metal during solidification of ingot; *or* cylindrical metal component used for conveying fluids.

pitch. Center-to-center distance of tube holes or stays; distance from thread crest to thread crest.

plus heads. Formed heads having pressure on the concave side.

porosity. In a weld, gas bubbles that become trapped as the weld metal solidifies.

postweld heat treatment. Any heat treatment applied after welding.

power boiler. Pressure vessel in which steam is generated at more than 15 psig for use external to itself.

PQR. Welding procedure qualification record.

preheat. Warming of the parent metals prior to welding. See PW-38.

proof test. Method of establishing the pressure rating for a component by pressure test. See PG-18.

PWHT. Postweld heat treatment.

quality (of steam). Measure of the dryness of saturated steam. Quality is the ratio of the mass of vapor to the total mass of water present.

quality factor. Number not greater than 1 that is multiplied by the allowable stress values tabulated in Section II, Part D, for castings. Depends on the degree of quality assurance.

quick-opening valve. Valve that may be opened by turning less than five full turns of its handle. Blowoff piping requires at least one slow-opening valve and one that may be slow or fast.

refractory. Firebrick and masonry work in a boiler furnace.

reheat superheater. Tubes arranged to permit steam to be exposed to heat after it has been passed through part of a turbine or otherwise partly used.

relief valve. Spring-loaded, pressure-operated valve characterized by gradual opening as pressure increases used to prevent overpressure of a hot water boiler.

relieving capacity. The amount of steam, in pounds per hour, that will pass through a safety valve at its set pressure.

remote level indicator. Device other than the gage glass for showing the water level in a boiler.

repair organization. Holder of National Board certificate of authorization for use of R stamp. Note the R stamp is not issued by ASME.

rerating. Increase in the maximum allowable working pressure of a boiler or pressure vessel. May also be an increase in operating temperature.

rimmed steel. Steel with high porosity and tendency for segregation produced by casting ingots with no chemical deoxidation.

Rockwell hardness. Penetration hardness test based on the depth of the impression left by a diamond or steel ball indenter under a specified load.

rupture disk. Device that may be used on organic fluid vaporizers between the safety valve and the vaporizer. They rupture within 5 percent of nominal pressure. See PVG-12.3.

safety relief valve. Spring-loaded, pressure-operated valve suitable for use as either a safety valve or a relief valve.

safety valve. Spring-loaded, pressure-operated valve characterized by pop opening used to prevent overpressure of a boiler. Used for steam or vapor service.

saturation. Condition of pressure and temperature that produces equilibrium between vapor and liquid phases of water, e.g., 212°F and 14.7 psia.

schedule number. In the nomenclature of pipe, the schedule number indicates the thickness class; for example, schedule 40 is standard and 80 is extra heavy.

Scotch boiler. A type of internally fired firetube boiler having one or more furnaces inside a water-containing shell and horizontal return firetubes.

seal weld. Weld attaching a tube to a tubesheet, intended only for prevention of leakage. The holding power comes from expanding the tube.

Section IX. Welding and brazing qualifications section of the ASME Code.

Section V. Nondestructive testing section of ASME Code.

semiellipsoidal. Type of formed head having the shape of an ellipse, the major axis of which is twice the minor axis. See PG-29.7 and 29.8.

semikilled steel. Partially deoxidized steel characterized by moderate porosity in the ingot and not much pipe. The ingot has a high usable yield.

setting. The structural support system for a boiler.

shell. The cylindrical portion of a pressure vessel. The vessel consists of a shell with heads on its ends.

slag inclusion. Nonmetallic inclusion in a weld, generally from the electrode coating or flux.

slow-opening valve. A valve requiring five or more full turns to go from fully closed to fully open. See B31.1, paragraph 122.1.7(C.8).

SNT-TC-1A. Training document published by ASNT; standard for qualification of NDT personnel. The 1995 Code references the 1984 edition of SNT-TC-1A.

standard pressure parts. See PG-11. Parts that may be used on a boiler without partial data reports and inspection in the shop of parts manufacturer.

stay. Device for supporting flat components exposed to pressure.

staybolts. Short stays either welded or threaded, running between two plates, as in a water leg.

steam trap. Device for removing condensate from steam.

steaming capacity. The amount of steam, in pounds per hour, that a boiler is capable of generating at its maximum firing rate.

stop valve. Valve in steam or feedwater lines for shutting off or turning on flow. Typically a stop valve is either fully open or fully closed.

strength weld. Weld joining a tube to a tubesheet designed to develop the full tensile strength of the tube. Also provides leak tightness. See UW-20.

stress relief. Heat treatment consisting of heating to a temperature and for a time sufficient to eliminate most residual stresses, especially from welding, followed by slow cooling.

superheated steam. Steam that is at a higher temperature than the saturation temperature for the pressure.

superheater. Device for producing superheated steam; typically a group of tubes exposed to heat through which steam is passed.

Synopsis. Book of Jurisdictions and laws by the Uniform Boiler & Pressure Vessel Laws Society.

syphon. Device on a steam pressure gage to protect the gage from the entry of live steam. Also called a pig tail. A loop in the line to the gage.

tack weld. Small temporary weld used to hold parts to be welded in proper position for welding.

tell-tale hole. In a staybolt, a hole in the end to emit steam if the staybolt becomes cracked—see PG-47.1; *or,* in a compensation pad or "saddle," a hole through the pad to permit testing of welds for leaks—see PW-15.3.

test cocks. See gage cocks.

test plates. See vessel test plates.

thermal expansion. Change in dimensions of parts as the temperature changes.

throat. Distance from the root of a fillet weld to its surface. For a fillet weld with equal legs and straight contour, the throat is 0.7 times the leg length.

through stays. Long rods running from one head to the other to support the heads in firetube boilers.

torispherical. Type of formed head with a central portion dished to a segment of a sphere, a corner or knuckle radius, and a straight cylindrical flange. See PG-29.

try cocks. See gage cocks.

tube. Cylindrical metal component used for the transfer of heat.

tube sheet. Flat plate or head into which tube ends are installed. Also that portion of a cylindrical drum into which the tubes are inserted.

UTS. Ultimate tensile strength; in a tensile test, the maximum load sustained by a sample divided by the original cross-sectional area.

vessel test plates. Weld sample plates required to be prepared and tested per PW-53 for shells and drums not made of P-1 materials.

waste heat boiler. Boiler in which the heat for generating the steam is derived from hot gases from a source outside the boiler, not from the combustion of fuel in the boiler.

water column. A tube or cylinder fastened to a drum or shell of a boiler to allow the water level to be established. See Figure PG-60.

water hammer. Shock to a pipe or vessel caused by impact of a slug of water striking a surface.

water leg. Area between two plates containing water to cool the sides of an internally fired furnace. See Figure PWT-12.2.

weld metal tension test. Required test on vessel plate in which a sample is cut from the plate such that the sample axis is the weld axis and the sample is all weld metal, end to end.

welder. The person performing manual or semiautomatic welding.

welding operator. Operator of an automatic welding process.

WPS. Welding procedure specification, often called the welding procedure; Written statement of essential and nonessential variables for making a weld.

wrapper sheet. Outside plate surrounding a combustion chamber.

yield point. That value of stress where a material stretches at no increase in load.

yoke. A type of A.C. or D.C. horseshoe magnet used in magnetic particle inspection; *or* part of the clamping device used to install manhole and handhole covers.

References

The American Society of Mechanical Engineers Boiler & Pressure Vessel Code, "Rules for Construction of Power Boilers," sec. 1, 1995.

———, "Rules for Construction of Power Boilers," Sec. I, 1992.

———, "Rules for Construction of Power Boilers," Sec. I, 1989.

———, "Rules for Construction of Power Boilers," Sec. I, 1959.

———, "Rules for Construction of Power Boilers," Sec. I, 1952.

———, "Rules for Construction of Power Boilers," Sec. I, 1949.

———, "Rules for Construction of Power Boilers," Sec. I, 1946.

———, "Materials Part D: Properties," Sec. II, 1995.

———, "Materials Part D: Properties," Sec. II, 1992.

———, "Materials Part A: Ferrous," Sec. II, 1989.

———, "Rules for Construction of Heating Boilers," Sec. IV, 1995.

———, "Rules for Construction of Heating Boilers," Sec. IV, 1992.

———, "Nondestructive Examination," Sec. V, 1995.

———, "Nondestructive Examination," Sec. V, 1992.

———, "Rules for Construction of Pressure Vessels," Sec. VIII, 1995.

———, "Rules for Construction of Pressure Vessels," Sec. VIII, 1992.

———, "Rules for Construction of Pressure Vessels," Sec. VIII, 1968.

———, "Welding and Brazing Qualifications," Sec. IX, 1995.

———, "Welding and Brazing Qualifications," Sec. IX, 1992.

ANSI/ASME, "Power Piping," B31.1, 1995.

———, "Power Piping," B31.1, 1992.

The Aultman & Taylor Machinery Co.: *"Cahall" Water Tube Steam Boilers,* 4th ed., 1897.

Babcock & Wilcox Company: *Steam: Its Generation and Use,* 40th ed., 1992.

———: *Steam: Its Generation and Use,* 37th ed., 1963.

———: *Steam Superheaters,* 2nd ed., 1923.

Bernstein, Martin D. and Lloyd W. Yoder: *Course Notes: ASME Professional Development Course, Section I,* 1995.

Carroll, Dyer E. and Dyer E. Carroll, Jr.: *Course Notes: National Board Examination Preparation Course,* 1995.

The Clark Reliance Corporation: *Catalog 500 Water Column Safety Systems.*

Combustion Engineering Company: *Operation & Maintenance of Main Boiler Units, U.S. Navy Escort Vessels, DE 51 Class,* 1942.

Consolidated Safety Valves: *Bulletin SV 6.*

Diamond Power Babcock & Wilcox: *Instruction Manual for the Diamond Model 1K Sootblower,* rev. ed., 1991.

———: *Instruction Manual for the Diamond Model G9B Sootblower,* rev. ed., 1988.

Diamond Power Specialty Corp.: *Instruction Manual for the Diamond Model 1K Sootblower,* rev. ed., 1969.

Elliot Company: *Basic Principles of Tube Expanding.*

Hawkins, N.: *Maxims and Instructions for the Boiler Room,* Audel & Co, New York.

The Hodge Boiler Works: *Steam Boilers,* 1903.

Industrial Code Bulletin No. 14: Rules as Amended Relating to the Construction, Installation & Maintenance of Steam Boilers, State of New York, Department of Labor, Board of Standards and Appeals, 1940.

Keenan and Keyes: *Thermodynamic Properties of Steam,* rev. ed., John Wiley & Sons, New York, 1936.

de Lorenzi, Otto (ed.): *Combustion Engineering: A Reference Book on Fuel Burning and Steam Generation,* 1st edition, Combustion Engineering Superheater, Inc., 1949.

Lyon, Frank and A. W. Hinds: *Marine & Naval Boilers,* United States Naval Institute, 1915.

Marks, Lionel S. (ed): *Steel Plates and Their Fabrication,* Lukens Steel Company, 1947.

Milton, J. H.: *Marine Steam Boilers,* 3rd ed., Newnes-Butterworths, London, 1970.

Mutual Boiler and Machinery Insurance Company: *Boiler & Machinery Insurance.*

The National Board Inspection Code, ANSI/NB 23, 1992.

National Board of Boiler and Pressure Vessel Inspectors: *NB 57: The National Board of Boiler and Pressure Vessel Inspectors Guide for ASME Code Sections I, IV, and VIII, Divisions 1 and 2, Reviews and the National Board "R" Certificate of Authorization,* 2nd ed., 1990.

———, *NB 215: Rules and Regulations,* 1994.

Ohio Boiler Inspection Law and Rules, The Industrial Commission of Ohio, 1947.

The Ohio Injector Company Catalog, no. 14.

Osbourne: *Modern Marine Engineer's Manual,* Cornell Maritime Press, New York, 1941.

Singer, Joseph G. (ed.): *Combustion Fossil Power: A Reference Book on Fuel Burning and Steam Generation,* 4th ed., Combustion Engineering, Inc., 1991.

Sothern: *"Verbal" Notes and Sketches,* 14th ed., vols. 1 and 2, James Munro & Co., Glasgow, 1942.

Spring, Harry M. and Anthony L. Kohan: *Boiler Operator's Guide,* 2nd ed., McGraw-Hill, New York, 1981.

Steam Boiler Rules, Form U, Commonwealth of Massachusetts, Department of Public Safety, Board of Boiler Rules, 1936 (amended 1947).

Uniform Boiler and Pressure Vessel Laws Society, Inc. *Synopsis of Boiler and Pressure Vessel Laws, Rules and Regulations,* rev. ed., 1994.

USS Tubular Products: *USS Standard Pipe and Line Pipe Catalog,* 1992.

Index of Sample Problems

Index

ABOUT THE AUTHOR

Dyer E. Carroll is founder and president of Carroll Engineers, Inc., a consulting firm based in Andover, Massachusetts that specializes in nondestructive testing, boiler and pressure vessel design, and forensic engineering. He also is a professor in the Fire Technology Program at Massachusetts Bay Community College and teaches the Engineering Inspectors' Training Program for the Lloyds Register Shipping Insurance Company. A Registered Professional Engineer and Licensed First Class Stationary Engineer in Massachusetts, Mr. Carroll hold commissions from the National Board of Boiler and Pressure Vessel Inspectors in seven states.

Dyer E. Carroll, Jr. is vice president and radiation safety officer of Carroll Engineers, Inc. He holds a Master of Science degree in Mechanical Engineering from Northeastern University and has extensive experience as a consulting and forensic engineer.